The Eye of the Master

The Eye of the Master

A Social History of Artificial Intelligence

Matteo Pasquinelli

London • New York

First published by Verso 2023
© Matteo Pasquinelli 2023

A version of chapter 4 was published for *Radical Philosophy* issue 2.06 in Winter 2019. An earlier version of chapter 8 was published for the Berkeley University journal *Qui Parle* in Spring 2021.

All rights reserved

The moral rights of the author have been asserted

3 5 7 9 10 8 6 4 2

Verso
UK: 6 Meard Street, London W1F 0EG
US: 388 Atlantic Avenue, Brooklyn, NY 11217
versobooks.com

Verso is the imprint of New Left Books

ISBN-13: 978-1-78873-006-8
ISBN-13: 978-1-78873-007-5 (UK EBK)
ISBN-13: 978-1-78873-008-2 (US EBK)

British Library Cataloguing in Publication Data
A catalogue record for this book is available from the British Library

Library of Congress Cataloging-in-Publication Data

Names: Pasquinelli, Matteo, author.
Title: The eye of the master : a social history of artificial intelligence / Matteo Pasquinelli.
Description: London ; New York : Verso, 2023. | Includes bibliographical references and index.
Identifiers: LCCN 2023020753 (print) | LCCN 2023020754 (ebook) | ISBN 9781788730068 (trade paperback) | ISBN 9781788730082 (ebook)
Subjects: LCSH: Automation—Social aspects—History. | Machinery in the workplace—Social aspects—History. | Artificial intelligence—Social aspects—History. | Employees—Effect of technological innovations on—History.
Classification: LCC HC79.A9 P37 2023 (print) | LCC HC79.A9 (ebook) | DDC 338/.064—dc23/eng/20230609
LC record available at https://lccn.loc.gov/2023020753
LC ebook record available at https://lccn.loc.gov/2023020754

Typeset in Minion by Hewer Text UK Ltd, Edinburgh
Printed and bound by CPI Group (UK) Ltd, Croydon CR0 4YY

Contents

List of Illustrations	vii
Introduction: AI as Division of Labour	1
1 The Material Tools of Algorithmic Thinking	23

Part I
THE INDUSTRIAL AGE — 49

2 Babbage and the Mechanisation of Mental Labour	51
3 The Machinery Question	77
4 The Origins of Marx's General Intellect	95
5 The Abstraction of Labour	121

Part II
THE INFORMATION AGE — 131

6 The Self-Organisation of the Cybernetic Mind	133
7 The Automation of Pattern Recognition	161

8 Hayek and the Epistemology of Connectionism   182
9 The Invention of the Perceptron   205

Conclusion: The Automation of General Intelligence   237

Acknowledgements   254
Index   257

Illustrations

1.1 Diagram of the Agnicayana fire altar. Frits Staal, 'Greek and Vedic Geometry', *Journal of Indian Philosophy* 27, no. 1 (1999): 111 (image rotated). 23

1.2 The reflective structure of abstraction. Peter Damerow, *Abstraction and Representation*, Berlin: Springer, 2013, 379. 37

1.3 Allegory of Arithmetic. Gregor Reisch, *Margarita Philosophica*, 1503. 42

2.1 Scheme for the implementation of de Prony's algorithm as division of labour. Lorraine Daston, 'Calculation and the Division of Labor, 1750–1950', *Bulletin of the German Historical Institute* 62 (Spring 2018): 11. 57

2.2 Design for the implementation of de Prony's algorithm into a mechanism. Charles Babbage, *On the Economy of Machinery and Manufactures*, London: Charles Knight, 1832, 161. 60

2.3 Babbage's Difference Engine. Charles Babbage, *Passages from the Life of a Philosopher*, London: Longman, Roberts & Green, 1864, front cover. 61

4.1 William Heath, 'The March of Intellect', ca. 1828, print, British Museum. 97

6.1 Sketch of artificial neurons. Warren McCulloch and Walter Pitts, 'A Logical Calculus of the Ideas Immanent in Nervous Activity', *Bulletin of Mathematical Biophysics* 5, no. 4 (1943): 105. 135

6.2 Operating arm of the universal constructor. John von Neumann, *Theory of Self-Reproducing Automata* (edited by Arthur Burks), Urbana, Il: University of Illinois Press, 1966, 371. 142

6.3 Diagram showing patterns of dappling and calculations. Alan Turing, ca. 1950. Sheet AMT/K3/8, Turing Archive, King's College Cambridge, particular. 145

7.1 Diagram of the superior colliculus of the midbrain. Warren McCulloch and Walter Pitts, 'How We Know Universals: The Perception of Auditory and Visual Forms', *Bulletin of Mathematical Biophysics* 9, no. 3 (1947): 141. 167

9.1 Examples of target classification. Albert Murray, 'Perceptron Applications in Photo Interpretation', *Photogrammetric Engineering* 27, no. 4 (1961): 633. 208

9.2 Examples of target classification. Albert Murray, 'Perceptron Applications in Photo Interpretation', *Photogrammetric Engineering* 27, no. 4 (1961): 634. 209

9.3 Mark I Perceptron. Frank Rosenblatt, *Principles of Neurodynamics: Perceptrons and the Theory of Brain Mechanisms*, Buffalo, NY: Cornell Aeronautical Laboratory, 1961, iii. 210

9.4 Diagram of the organisation of the Mark I Perceptron. Frank Rosenblatt, *Mark I Perceptron Operators' Manual*. Buffalo, NY: Cornell Aeronautical Laboratory, 1960, 4. 211

9.5 Sketch of the simple perceptron. Frank Rosenblatt, 'Audio Signal Pattern Perception Device', US patent US3287649A, 1963 (expired 1983). 214

Introduction: AI as Division of Labour

The special skill of each individual machine-operator, who has now been deprived of all significance, vanishes as an infinitesimal quantity in the face of the science, the gigantic natural forces, and the mass of the social labour embodied in the system of machinery, which, together with these three forces, constitutes the power of the 'master'.

Karl Marx, Capital, 1867[1]

All human beings are intellectuals . . . although one can speak of intellectuals, one cannot speak of non-intellectuals, because non-intellectuals do not exist . . . There is no human activity from which every form of intellectual participation can be excluded: Homo faber *cannot be separated from* homo sapiens.

Antonio Gramsci, The Prison Notebooks [1932][2]

In the twentieth century, few would have ever defined a truck driver as a 'cognitive worker', an intellectual. In the early twenty-first, however, the application of artificial intelligence (AI) in self-driving vehicles, among other artefacts, has changed the perception of manual skills such

---

1 Karl Marx, *A Critique of Political Economy*, vol. 1, trans. Ben Fowkes, London: Penguin, 1981, 549. Marx remarked also: 'The "master" can sing quite another song, when he is threatened with the loss of his "living" automaton' (ibid., 544).

2 Antonio Gramsci, *Selections from the Prison Notebooks*, New York: International Publishers, 1971, 9 (translation modified).

as driving, revealing how the most valuable component of work in general has never been just manual, but has always been cognitive and cooperative as well. Thanks to AI research – we must acknowledge it – truck drivers have reached the pantheon of intelligentsia. It is a paradox – a bitter political revelation – that the most zealous development of automation has shown how much 'intelligence' is expressed by activities and jobs that are usually deemed manual and unskilled, an aspect that has often been neglected by labour organisation as much as critical theory. In fact, in the current digital age, only a few sociologists, such as Richard Sennett, have taken the trouble to emphasise that 'making is thinking', a dimension that the historians of science such as Lissa Roberts and Simon Schaffer have captured in the elegant image of the 'mindful hand' – a hand that, in the workshop of the Renaissance as much as in those of the industrial age, has not only expressed muscular strength but also inspired design, inventions, and scientific breakthroughs.[3] If there is a denial of the intelligence of manual labour and social activities today, that seems to be a symptom also of the overgrowth of the digital sphere and the dematerialisation of human activities, which have contributed to the aura of mystery that has been eventually constructed around AI.

What is AI? A dominant view describes it as the quest 'to solve intelligence' – a solution supposedly to be found in the secret logic of the mind or in the deep physiology of the brain, such as in its complex neural networks. In this book I argue, to the contrary, that the inner code of AI is constituted not by the imitation of biological intelligence but by the *intelligence of labour and social relations*. Today, it should be evident that AI is a project to capture the knowledge expressed through individual and collective behaviours and encode it into algorithmic models to automate the most diverse tasks: from image recognition and object manipulation to language translation and decision-making. As in a typical effect of ideology, the 'solution' to the enigma of AI is in front of our eyes, but nobody can see it – nor does anybody want to.

---

3 Richard Sennet, *The Craftsman*, New Haven, CT: Yale University Press, 2008; Lissa Roberts and Simon Schaffer (eds), *The Mindful Hand: Inquiry and Invention from the Late Renaissance to Early Industrialisation*, Chicago: University of Chicago Press, 2007. On the role of touch in the digital age, see Rebekka Ladewig and Henning Schmidgen (eds), *Body and Society* 28, nos. 1–2, special issue, 'Symmetries of Touch: Reconsidering Tactility in the Age of Ubiquitous Computing' (2012).

Let us return to the contested project of the self-driving car. What kind of work does a driver perform? And to what extent can AI automate such an activity? With a considerable degree of approximation and hazard, a self-driving vehicle is designed to imitate all the micro-decisions that a driver makes on a busy road.[4] Its artificial neural networks 'learn' the correlations between the visual perception of the environment and the mechanical control of the vehicle (steering, accelerating, braking) together with ethical decisions to be taken in a few milliseconds in case of danger. Driving requires high cognitive skills that cannot be left to improvisation, but also rapid problem-solving that is possible only thanks to habit and training that are not completely conscious. Driving remains essentially a social and cooperative activity, which follows both codified rules (with legal constraints) and spontaneous ones, including a tacit cultural code which is different in each locality. It is deemed difficult to encode such a complex activity, and even the entrepreneur Elon Musk has admitted, after not a few fatal accidents of Tesla cars, that 'generalized self-driving is a hard problem'.[5] In all its problematic aspects, however, the industrial project of self-driving vehicles has made clear that the task of driving is not merely 'mechanical'. If the skill of driving can be translated into an algorithmic model to begin with, it is because driving is a logical activity – because, ultimately, *all labour is logic.*[6]

What, then, is the relationship between labour, rules, and automation, i.e., the invention of new technologies? This entanglement is the core problem of AI which this book seeks to explore. But this is not a completely new perspective for framing AI. The historian of science Lorraine Daston, for example, has already illustrated this problem in the great calculation projects of the Enlightenment that preceded automatic computation. In the late eighteenth century, in order to produce the lengthy logarithmic tables necessary for the modernisation of revolutionary France, the

---

4 In the early 1960s, Romano Alquati defined 'information' as the innovative microdecisions that workers take along the production process (see chapter 5 in this book). See also Matteo Pasquinelli, 'Italian Operaismo and the Information Machine', *Theory, Culture and Society* 32, no. 3 (2015): 49–68; Florian Sprenger, 'Microdecisions and Autonomy in Self-Driving Cars: Virtual Probabilities', *AI and Society* (2020): 1–16.

5 Andrew J. Hawkins, 'Elon Musk Just Now Realizing That Self-Driving Cars Are a "Hard Problem"', theverge.com, 5 July 2021.

6 See also 'A Manifesto', *Logic Magazine*, issue 1, March 2017.

mathematician Gaspard de Prony had the idea to apply the industrial method of the division of labour (canonised by Adam Smith in *The Wealth of Nations*) to hand calculation.[7] For this purpose, de Prony arranged a *social algorithm* – a hierarchical organisation of three groups of clerks which divided the toil and each performed one part of the long calculation, eventually composing the final results. A few years later, in industrial England, Charles Babbage adopted the intuition of the division of labour as the internal principle of the Difference Engine, designing in this way the first prototype of the modern computer. Babbage, importantly, understood that the division of labour was not only a principle to design machines but also to compute the costs of production (what has been known since then as the 'Babbage principle').

In the industrial age, the supervision of the division of labour used to be the task of the factory's master.[8] The eye of the master, in workshops and also in camps and plantations, had long supervised and disciplined workers, drawing the plans of assembly lines as well as the shifts of forced labour. Before industrial machines were invented, urban sweatshops and colonial estates were already 'mechanical' in their regime of body discipline and visuality.[9] As the philosopher Michel Foucault illustrated, the imposition of such disciplinary techniques – based on the segmentation of time, space, and relations – prepared the terrain for the capitalist regime of labour exploitation.[10] In parallel, the rationalist view

---

7 Lorraine Daston, 'Calculation and the Division of Labor, 1750–1950', *Bulletin of the German Historical Institute* 62 (Spring 2018): 13.

8 Friedrich Engels contested 'the eye of the master' and argued that in larger enterprises and infrastructures of the time, such as the railway companies, workers possess a greater overview of the production process: 'The economical development of our actual society tends more and more to concentrate, to socialise production into immense establishments which cannot any longer be managed by single capitalists. All the trash of "the eye of the master", and the wonders it does, turns into sheer nonsense as soon as an undertaking reaches a certain size. Imagine "the eye of the master" of the London and North Western Railway! But what the master cannot do the workman, the wages-paid servants of the Company, can do, and do it successfully. Thus the capitalist can no longer lay claim to his profits as "wages of supervision", as he supervises nothing.' Friedrich Engels, 'Social Classes: Necessary and Superfluous', *Labour Standard*, 6 August 1881.

9 On the regime of visuality in the slave plantation, see Nicholas Mirzoeff, *The Right to Look: A Counterhistory of Visuality*, Durham, NC: Duke University Press, 2011.

10 Michel Foucault, *Discipline and Punish: The Birth of the Prison*, New York: Pantheon Books, 1977.

Introduction: AI as Division of Labour

of the world helped to further describe the movement of the human body in detail and draft its mechanisation. Historian Sigfried Giedion detailed this process in his famous volume *Mechanisation Takes Command*. According to Giedion, mechanisation begins 'with the concept of Movement', then it replaces handicraft, and, finally, its full development is 'the assembly line, wherein the entire factory is consolidated into a synchronous organism'.[11]

This mechanical mentality culminated in Taylorism – a system of 'scientific management' that sought to economize workers' movements down to the finest detail. Indeed, as the political economist Harry Braverman once noted, 'Taylor understood the Babbage principle better than anyone of his time, and it was always uppermost in his calculations.'[12] In order to surveil the worker's smallest gesture, the Taylorist system even acquired cinematographic eyes: the factory's master became a sort of movie director who filmed workers in order to measure and optimize their productivity, somehow realising what media scholar Jon Beller has termed the 'cinematic mode of production'.[13] Taylorism prompted the discipline of 'time and motion study' which was pursued, in the same years, by both the Soviet revolutionary Aleksei Gastev and the US engineers Frank and Lillian Gilbreth, who introduced similar photographic techniques such as, respectively, the *cyclogram*, and *chronocyclegraph*.[14] This book follows these analytical studies of the labour process through

---

11 Sigfried Giedion, *Mechanization Takes Command: A Contribution to Anonymous History*, Oxford: Oxford University Press, 1948, 5.

12 Harry Braverman, *Labor and Monopoly Capital: The Degradation of Work in the Twentieth Century*, New York: Monthly Review Press, 1974, 117. See Frederick Winslow Taylor, *The Principles of Scientific Management*, New York: Harper & Brothers, 1911.

13 Jonathan Beller, *The Cinematic Mode of Production: Attention Economy and the Society of the Spectacle*, Lebanon, IN: University Press of New England, 2006.

14 See Charles S. Maier, 'Between Taylorism and Technocracy: European Ideologies and the Vision of Industrial Productivity in the 1920s', *Journal of Contemporary History* 5, no. 2 (1970): 27–61; Rolf Hellebust, 'Aleksei Gastev and the Metallization of the Revolutionary Body', *Slavic Review* 56, no. 3 (1997): 500–18; Ana Hedberg Olenina, *Psychomotor Aesthetics: Movement and Affect in Modern Literature and Film*, Oxford: Oxford University Press, 2020; Nicolás Salazar Sutil, *Motion and Representation: The Language of Human Movement*, Cambridge, MA: MIT Press, 2015; Elspeth Brown, *The Corporate Eye: Photography and the Rationalization of American Commercial Culture, 1884–1929*, Baltimore, MD: Johns Hopkins University Press, 2008; Daniel Nelson (ed.), *A Mental Revolution: Scientific Management Since Taylor*, Columbus, OH: Ohio State University Press, 1992.

the industrial age up to the rise of AI, aiming to show how the 'intelligence' of technological innovation has often originated from the imitation of these abstract diagrams of human praxis and collective behaviours.

When industrial machines such as looms and lathes were invented, in fact, it was not thanks to the solitary genius of an engineer but through the imitation of the collective diagram of labour: by capturing the patterns of hand movements and tools, the subdued creativity of workers' know-how, and turning them into mechanical artefacts. Following this theory of invention, which was already shared by Smith, Babbage, and Marx in the nineteenth century, this book argues that the most sophisticated 'intelligent' machines have also emerged by imitating the outline of the collective division of labour. In the course of this book, this theory of technological development is renamed the *labour theory of automation*, or *labour theory of the machine*, which I then extend to the study of contemporary AI and generalise into a *labour theory of machine intelligence*.[15]

Already for Marx, the master was no longer an individual but, as mentioned in the opening quote to this introduction, an integrated power made up of 'the science, the gigantic natural forces, and the mass of the social labour embodied in the system of machinery'. After the expansion of 'the division of labour in society', as Émile Durkheim recorded at the end of the nineteenth century, the eye of the master evolved as well into new technologies of control such as statistics and the global 'operations of capital' (to use Sandro Mezzadra and Brett Neilson's apt phrase).[16] Since the end of the twentieth century, then, the management of labour has turned all of society into a 'digital factory' and has taken the form of the software of search engines, online maps, messaging apps, social networks, gig-economy platforms, mobility services, and ultimately AI algorithms, which have been increasingly used to automate all the abovementioned services.[17] It is not difficult to

---

15 The expressions 'labour theory of automation' and 'labour theory of the machine' are constructed on similar ones that are used in political economy such as the 'labour theory of value'.

16 Émile Durkheim, *The Division of Labor in Society*, New York: Free Press, 1984 [1893]; Sandro Mezzadra and Brett Neilson, *The Politics of Operations: Excavating Contemporary Capitalism*, Durham, NC: Duke University Press, 2019, 55.

17 See Moritz Altenried, *The Digital Factory: The Human Labor of Automation*, Chicago: University of Chicago Press, 2021.

Introduction: AI as Division of Labour

see AI nowadays as a further centralisation of digital society and the orchestration of the division of labour throughout society.

The thesis that the design of computation and 'intelligent machines' follow the schema of the division of labour is not heretical but receives confirmation from the founding theories of computer science, which have inherited a subtext of colonial fantasy and class division from the industrial age. The celebrated genius of automated computation Alan Turing, for instance, himself reiterated a hierarchical and authoritarian mode of thinking. In a 1947 lecture, Turing envisioned the Automatic Computing Engine (ACE), one of the first digital computers, as a centralised apparatus that orchestrated its operations as a hierarchy of master and servant roles:

> Roughly speaking those who work in connection with the ACE will be divided into its masters and its servants. Its masters will plan out instruction tables for it, thinking up deeper and deeper ways of using it. Its servants will feed it with cards as it calls for them. They will put right any parts that go wrong. They will assemble data that it requires. In fact the servants will take the place of limbs. As time goes on the calculator itself will take over the functions both of masters and of servants. The servants will be replaced by mechanical and electrical limbs and sense organs. One might for instance provide curve followers to enable data to be taken direct from curves instead of having girls read off values and punch them on cards. The masters are liable to get replaced because as soon as any technique becomes at all stereotyped it becomes possible to devise a system of instruction tables which will enable the electronic computer to do it for itself. It may happen however that the masters will refuse to do this. They may be unwilling to let their jobs be stolen from them in this way. In that case they would surround the whole of their work with mystery and make excuses, couched in well chosen gibberish, whenever any dangerous suggestions were made. I think that a reaction of this kind is a very real danger.[18]

The prose of the young Turing, in dividing computing tasks between 'masters', 'servants', and 'girls' is fierce. It is reminiscent of Andrew Ure's

---

18 Alan Turing, 'Lecture on the Automatic Computing Engine' (1947), in *The Essential Turing*, ed. B. Jack Copeland, London: Clarendon Press, 2004, 392.

gothic depictions of the industrial factory in the Victorian age as 'a vast automaton, composed of various mechanical and intellectual organs, acting in uninterrupted concern for the production of a common object, all of them being subordinated to a self-regulated moving force'.[19] Similarly, Turing imagined an intelligent automaton that in the future would be able to reprogram itself and replace both masters and servants. Turing's vision is contradicted today by the army of 'ghost workers' from the Global South, who, as Mary Gray and Siddharth Suri have documented, are removed from sight to let the show of machine autonomy go on.[20] Paradoxically for Turing, AI came to replace mostly masters, that is managers, rather than servants – workers are needed (and always will be) to produce data and value for the voracious pipelines of AI and its global monopolies, and, on the other hand, to provide the maintenance of such a mega-machine under the form of content filtering, security checks, evaluation and non-stop optimisation. As gender studies scholars Neda Atanasoski and Kalindi Vora have pointed out, the dreams of full automation and AI such as Turing's are not neutral but are historically grounded on the 'surrogate humanity' of enslaved servants, proletarians, and women that have made possible, through their invisible labour, the universalistic ideal of the free and autonomous (white) subject.[21]

The many histories of AI

Writing a history of AI in the current predicament means reckoning with a vast ideological construct: among the ranks of Silicon Valley companies and also hi-tech universities, propaganda about the almighty power of AI is the norm and sometimes even repeats the folklore of

---

19 Andrew Ure, *The Philosophy of Manufactures*, London: Charles Knight, 1835, 13–14.

20 Mary Gray and Siddharth Suri, *Ghost Work: How to Stop Silicon Valley from Building a New Global Underclass*, New York: Houghton Mifflin Harcourt, 2019. See also Lilly Irani, 'The Cultural Work of Microwork', *New Media and Society* 17, no. 5, 2013, 720–39; Lilly Irani and Michael Six Silberman, 'Turkopticon: Interrupting Worker Invisibility in Amazon Mechanical Turk', Proceedings of CHI 2013, 28 April–2 May 2013.

21 Neda Atanasoski and Vora Kalindi, *Surrogate Humanity: Race, Robots, and the Politics of Technological Futures*, Durham, NC: Duke University Press, 2019.

machines achieving 'superhuman intelligence' and 'self-awareness'. This folklore is well exemplified by apocalyptic *Terminator* narratives, in which AI systems would achieve technological singularity and pose an 'existential risk' to the survival of the human species on this planet, as the futurologist Nick Bostrom, among others, professes.[22] Mythologies of technological autonomy and machine intelligence are nothing new: since the industrial age, they have existed to mystify the role of workers and subaltern classes.[23] As Schaffer has remarked, while describing the cult of automata in Babbage's age, 'To make machines look intelligent it was necessary that the sources of their power, the labour force which surrounded and ran them, be rendered invisible.'[24]

Speculative narratives aside, which never go into sufficient technical detail to clarify which kind of algorithms would actually execute 'superintelligence', one also finds today numerous *technical histories* of AI that, on the other hand, promise to explain its complex algorithms.[25] These technical overviews often voice corporate expectations for a 'master algorithm' that would solve all tasks of perception and cognition at a prodigious rate of information compression (because this is the very unromantic metrics by which 'intelligent' systems are ultimately assessed).[26] Once again, these readings rarely consider the historical contexts and social implications of automation, and draw a linear history of mathematical achievements which reinforces technological

---

22 Nick Bostrom, *Superintelligence: Paths, Dangers, Strategies*, Oxford: Oxford University Press, 2014.

23 Simon Schaffer, 'Babbage's Dancer and the Impresarios of Mechanism', in *Cultural Babbage: Technology, Time and Invention*, ed. Francis Spufford and Jenny Uglow, London: Faber & Faber, 1996; Simon Schaffer, 'Enlightened Automata', in *The Sciences in Enlightened Europe*, ed. William Clark, Jan Golinski, and Simon Schaffer, Chicago: University of Chicago Press, 1999; Elly Rachel Truitt, *Medieval Robots: Mechanism, Magic, Nature, and Art*, Philadelphia: University of Pennsylvania Press, 2015; Adelheid Voskuhl, *Androids in the Enlightenment: Mechanics, Artisans, and Cultures of the Self*, Chicago: University of Chicago Press, 2013.

24 Simon Schaffer, 'Babbage's Intelligence: Calculating Engines and the Factory System', *Critical Inquiry* 21, no. 1 (1994): 204. See also Bernard Geoghegan, 'Orientalism and Informatics: The Alterity in Artificial Intelligence, from the Chess-Playing Turk to Amazon's Mechanical Turk', *Ex-Position* 43 (June 2020): 45–90.

25 See, for instance, Nils Nilsson, *The Quest for Artificial Intelligence: A History of Ideas and Achievements*, Cambridge: Cambridge University Press, 2010.

26 For an example of corporate agenda see Pedro Domingos, *The Master Algorithm: How the Quest for the Ultimate Learning Machine Will Remake Our World*, London: Penguin, 2017.

determinism.[27] Within these technical histories of AI, one should also include cognitive science, as a considerable part of this field actually developed under the influence of computer science. Margaret Boden's monumental two-volume *Mind as Machine* (2006) remains probably one of the most detailed histories of AI as cognitive science, showing the complexity of this genealogy, however, without ideological fervour.

Resisting such narrow technical perspectives, a growing number of authors have started addressing the social implications of AI from the standpoint of workers, communities, minorities, and society as a whole. These authors question the virtuosity of algorithms that claim to be 'intelligent' while in fact amplifying inequalities, perpetuating gender and racial biases, and consolidating new form of knowledge extractivism. Thanks to books such as Cathy O'Neil's *Weapons of Math Destruction* (2016), Safiya Noble's *Algorithms of Oppression* (2018), Ruha Benjamin's *Race after Technology* (2019), and Wendy Chun's *Discriminating Data* (2021), among many others, the new field of critical AI studies is growing.[28] This novel scholarship builds upon older investigations of AI, cybernetics, and Cold War rationality from previous decades, among which should be included Alison Adam's *Artificial Knowing* (1998), Philip Agre's *Computation and Human Experience* (1997), Paul Edwards's *The Closed World* (1996), Joseph Weizenbaum's *Computer Power and Human Reason* (1976), and Hubert Dreyfus's paper for the Rand Corporation 'Alchemy and Artificial Intelligence' (1965), which is usually considered the first philosophical critique of AI.[29]

---

27 On the idea of autonomous technology, see the classic Langdon Winner, *Autonomous Technology: Technics-Out-of-Control as a Theme in Political Thought*, Cambridge, MA: MIT Press, 1977.

28 Cathy O'Neil, *Weapons of Math Destruction: How Big Data Increases Inequality and Threatens Democracy*, New York: Crown Publishers, 2016; Safiya Umoja Noble, *Algorithms of Oppression: How Search Engines Reinforce Racism*, New York: New York University Press, 2018; Ruha Benjamin, *Race after Technology: Abolitionist Tools for the New Jim Code*, Cambridge: Polity, 2019; Wendy Hui Kyong Chun, *Discriminating Data: Correlation, Neighborhoods, and the New Politics of Recognition*, Cambridge, MA: MIT Press, 2021. AI studies are not growing without internal complications. Yarden Katz has noted how sometimes 'critical AI experts use their position to reinforce white supremacy with a progressive face . . . with language appropriated from radical social movements'. Yarden Katz, *Artificial Whiteness*, New York: Columbia University Press, 2020, 128.

29 For a systematic overview of critical AI studies, see: University of Cambridge, Department of History and Philosophy of Science, Mellon Sawyer Seminar 'Histories of AI: A Genealogy of Power', May 2020–July 2021, ai.hps.cam.ac.uk.

Within the expanding landscape of critical works, this book's concern is to illuminate the social genealogy of AI and, importantly, the standpoint – the social classes – from which AI has been pursued as a vision of the world and epistemology. Different social groups and configurations of power have shaped information technologies and AI in the past century. Rather than on the 'shoulders of giants', as the saying goes, it could be said that the early paradigms of mechanical thinking and late machine intelligence have been developed, in different times and ways, 'on the shoulders' of merchants, soldiers, commanders, bureaucrats, spies, industrialists, managers, and workers.[30] In all these genealogies, the automation of labour has been the key factor, but this aspect is often neglected by a historiography of technology that privileges science's point of view 'from above'.

A common approach, for instance, links quite deterministically the rise of cybernetics, digital computation, and AI to abundant funding from the US military during World War II and in the Cold War period.[31] Yet recent studies have clarified that the archipelago of such 'war rationality' was quite unstable and cultivated paradigms such as game theory and linear programming that were also key in modelling the arms race and military logistics.[32] The influence of state apparatuses on information technologies, anyhow, started well before the military acceleration of World War II: the automation of information retrieval and statistical analysis dates back to the need to mechanise public bureaucracy and government work, at least since the 1890 United States census that introduced the Hollerith machine to process punched cards. The 'government machine' (as Jon Agar has called it) anticipated the rise of the large data centres of the digital age, which have been, as is notorious, not just the business of internet companies but also of intelligence agencies, as the mathematician Chris Wiggins and historian Matthew L. Jones have detailed.[33] In short, for more than a hundred years, it has

---

30 See Richard Hadden, *On the Shoulders of Merchants: Exchange and the Mathematical Conception of Nature in Early Modern Europe*, Albany, NY: State University of New York Press, 1994.

31 About the 'military a-priori' in the history of computation, see Geoffrey Winthrop-Young, 'Drill and Distraction in the Yellow Submarine: On the Dominance of War in Friedrich Kittler's Media Theory', *Critical Inquiry* 28, no. 4 (2002): 825–54.

32 See Paul Erickson et al., *How Reason Almost Lost Its Mind: The Strange Career of Cold War Rationality*, Chicago: University of Chicago Press, 2013.

33 Jon Agar, *The Government Machine: A Revolutionary History of the Computer*,

always been the accumulation of 'big data' about society and its behaviours that prompted the development of information technologies, from Hollerith's tabulator to machine learning itself.[34]

In summary, AI represents the continuation of data analytics techniques first supported by state bureaus, secretly cultivated by intelligence agencies, and ultimately consolidated by internet companies into a planetary business of surveillance and forecasting. This reading, however, is once again a history 'from above' that focuses on only the techniques of control and rarely the subjects on whom this control is exercised. The targets of this power (or 'surveillance capitalism' in Shoshana Zuboff's definition) are usually described not as actors possessing autonomy and 'intelligence' on their own but as passive subjects of measurement and control. This is a problem of critical theory in general and critical AI studies in particular: although these studies are concerned about the impact of AI on society, they often overlook the role of collective knowledge and labour as the primary source of the very 'intelligence' that AI comes to extract, encode, and commodify. Moreover, these studies often fail to see the contribution of social forms and forces to the key stages of technological invention and development. A true critical intervention should challenge this hegemonic position of AI as the unique 'master' of collective intelligence. The Italian philosopher Antonio Gramsci once argued against the hierarchies of education that 'all human beings are intellectuals': in a similar way, this book aims at rediscovering the centrality of the social intelligence that informs and empowers AI. It also contends – in a more radical thesis – that such social intelligence shapes the very design of AI algorithms from within.

This book is intended as an incursion into both the technical and social histories of AI, integrating these approaches into a *sociotechnical history* that may identify also the economic and political factors that influenced its inner logic. Rather than siding with a conventional *social constructivism* and going beyond the pioneering insights of *social informatics*, it tries to extend to the field of AI the method of *historical epistemology* – one propagated in the history of science, in a different way, by

---

Cambridge, MA: MIT Press, 2003; see Chris Wiggins and Matthew L. Jones, *How Data Happened: A History from the Age of Reason to the Age of Algorithms*, New York: W. W. Norton, 2023.

34 See Yarden Katz, 'Manufacturing an Artificial Intelligence Revolution', *SSRN Electronic Journal* (November 2017).

Boris Hessen, Henryk Grossmann, Georges Canguilhem, and Gaston Bachelard, and more recently by the work of the Max Planck Institute for the History of Science in Berlin, among other initiatives.[35] Where social constructivism generically emphasises the influence of external factors on science and technology, historical epistemology is concerned with the dialectical unfolding of social praxis, instruments of labour, and scientific abstractions within a global economic dynamics. This book attempts to study AI and *algorithmic thinking* in a similar way that historical and political epistemology has studied, in the modern age, the rise of *mechanical thinking* and scientific abstractions in relation to socio-economic developments.[36]

In this respect, over the past decades, a *political epistemology* of science and technology has also been strongly pursued by feminist theorists such as Hilary Rose, Sandra Harding, Evelyn Fox Keller, and Silvia Federici, among others. These authors have convincingly explained the rise of modern rationality and mechanical thinking (to which AI also belongs) in relation to the transformation of women's body, and the collective body in general, into a productive and docile machine.[37] In the traditions of political epistemology, we should also consider the labour process analysis that was initiated by Braverman's *Labor and Monopoly Capital* (1974) and the workers' inquiries of Italian *operaismo*, which Romano Alquati, for instance, conducted at the Olivetti computer

---

35 For a critique of social constructivism, see Langdon Winner, 'Upon Opening the Black Box and Finding It Empty: Social Constructivism and the Philosophy of Technology', *Science, Technology, and Human Values* 18, no. 3 (1993): 362–78. For an overview of historical epistemology, see Jürgen Renn, *The Evolution of Knowledge: Rethinking Science for the Anthropocene*, Princeton, NJ: Princeton University Press, 2020; Pietro Daniel Omodeo, *Political Epistemology: The Problem of Ideology in Science Studies*, Berlin: Springer, 2019; Henning Schmidgen, 'History of Science', in *The Routledge Companion to Literature and Science*, ed. Bruce Clarke and Manuela Rossini, London: Routledge, 2011.

36 See Peter Damerow et al., *Exploring the Limits of Preclassical Mechanics: A Study of Conceptual Development in Early Modern Science*, 2nd ed., New York: Springer, 2004; Matthias Schemmel, *Historical Epistemology of Space: From Primate Cognition to Spacetime Physics*, New York: Springer, 2015. For the notion of number, see chapter 1 in this book.

37 Hilary Rose and Steven Rose (eds), *The Radicalisation of Science*, London: Macmillan, 1976; Sandra Harding, *The Science Question in Feminism*, Ithaca, NY: Cornell University Press, 1986; Evelyn Fox Keller, *Reflections on Gender and Science*, New Haven, CT: Yale University Press, 1985; Silvia Federici, *Caliban and the Witch: Women, the Body, and Primitive Accumulation*, New York: Autonomedia, 2004.

factory in Ivrea as early as 1960.[38] Braverman and Alquati pioneered influential works that first showed how Babbage's automated computation projects in the nineteenth century as much as cybernetics in the twentieth were inherently related to the sphere of labour and its organisation.

The automation of cognition as pattern recognition

The translation of a labour process into a logical procedure and subsequently into a technical artefact is rarely straightforward and flawless; it often displays instead a spurious and experimental character. In this sense, the title *The Eye of the Master* contains not only a political but also a technical analogy. It signals, somewhat ironically, the ambivalence of the current paradigm of AI – deep learning – which emerged not from theories of cognition, as some may believe, but from contested experiments to automate the labour of perception, or pattern recognition.[39] Deep learning has evolved from the extension of 1950s techniques of visual pattern recognition to non-visual data, which now include text, audio, video, and behavioural data of the most diverse origins. The rise of deep learning dates to 2012, when the convolutional neural network AlexNet won the ImageNet computer vision competition. Since then, the term 'AI' has come to define by convention the paradigm of artificial neural networks which, in the 1950s, it must be noted, was actually its rival (an example of the controversies that characterise the 'rationality' of AI).[40] Stuart and Hubert Dreyfus illuminated this schism in their

---

38 See chapter 5. For a novel approach to workers' inquiry, see Jamie Woodcock, 'Towards a Digital Workerism: Workers' Inquiry, Methods, and Technologies', *Nanoethics* 15 (2021): 87–98.

39 'Master' and 'pattern' share a common political etymology. The English term 'pattern' comes from the French *patron* and the Latin *patronus*. Both have the same root of the English 'paternal' and 'father', that is the Latin *pater*. The Latin *patronus* means also protector, also in relation to servants. The French *patron* has the meaning of leader, boss, or head of a community, which, in patriarchal contexts, implies a model to follow.

40 AlexNet was a next-generation convolutional neural network named after Geoffrey Hinton's student Alex Krizhevsky. By convention, the following paper marks the beginning of the deep learning era: Alex Krizhevsky, Ilya Sutskever, and Geoffrey Hinton, 'Imagenet Classification with Deep Convolutional Neural Networks', *Advances in Neural Information Processing Systems* 25 (2012): 1097–105. See also Dominique Cardon, Jean-Philippe Cointet, and Antoine Mazières, 'Neurons Spike Back: The

1988 essay 'Making a Mind versus Modeling the Brain', in which they outlined the two lineages of AI – symbolic and connectionist – that, based on different logical postulates, have also followed different destinies.[41]

Symbolic AI is the lineage that is associated with the 1956 Dartmouth workshop for which John McCarthy coined the questionable term 'artificial intelligence'.[42] Its key applications have been the Logic Theorist and General Problem Solver – and the array of expert systems and inference engines in general – which were proven trivial and prone to combinatorial explosion. Connectionism, on the other hand, is the lineage of artificial neural networks pioneered by Frank Rosenblatt's invention of the 'perceptron' in 1957, which unfolded into convolutional neural networks in the late 1980s and, eventually, launched the deep learning architecture that has prevailed since the 2010s.

The two lineages pursue different kinds of logic and epistemology. The former professes that intelligence is a representation of the world (*knowing-that*) which can be formalised into propositions and, therefore, mechanised following *deductive logic*. The latter, in contrast, argues that intelligence is experience of the world (*knowing-how*) which can be implemented into approximate models constructed according to *inductive logic*. Pace corporate propaganda and computationalist philosophies of the mind, neither of these two paradigms has managed to fully imitate human intelligence. Machine learning and deep artificial neural networks, however, due to their resolution in rendering multidimensional data, have proven quite successful in techniques of pattern recognition and, therefore, in the automation of numerous tasks. Against a tradition which repeats the overly celebrated saga of the Dartmouth workshop, this book highlights the origins of artificial neural networks, connectionism, and machine learning as a more compelling history of AI about which, especially regarding Rosenblatt's work, critical and exhaustive literature is still missing.

Invention of Inductive Machines and the Artificial Intelligence Controversy', trans. Elizabeth Libbrecht, *Réseaux* 211, no. 5 (2018): 173–220.

41 Hubert Dreyfus and Stuart Dreyfus, 'Making a Mind versus Modeling the Brain: Artificial Intelligence Back at a Branchpoint', *Daedalus* 117, no. 1 (1988): 15–43.

42 John McCarthy et al., 'A Proposal for the Dartmouth Summer Research Project on Artificial Intelligence', 31 August 1955, reprinted in *AI Magazine* 27, no. 4 (2006).

Structure of the book

The book is divided into three sections: a methodological and introductory first chapter and two main historical parts on the industrial and information ages respectively. This book does not pursue, however, a linear history of technology and automation. Rather, each chapter can be read as an independent 'workshop' for the study of algorithmic practices and machine intelligence.

Chapter 1 moves from the need to clarify, before anything else, the central notion of computation: the algorithm. What is an algorithm? In computer science, it can be defined as a finite procedure of step-by-step instructions to turn an input into an output making the best use of the given resources. The chapter challenges this purely technical definition of the algorithm and argues for a materialist critique that may recognise its economic and social roots. After all, as with other abstract notions, such as number or mechanism, the algorithm has a long history; the mathematician Jean-Luc Chabert finds that 'algorithms have existed since the beginning of time and existed long before a special word was coined to describe them'.[43] By excavating the social mathematics of the ancient Hindu ritual Agnicayana, the chapter argues that algorithmic thinking and practices have belonged to all civilisations, not only to the metalanguage of Western computer science. Against mathematical and philosophical intuitionism, which believes in the full independence of mental constructs, the chapter stresses that algorithmic thinking emerged as a *material abstraction*, through the interaction of mind with tools, in order to change the world and solve mostly economic and social problems. Deliberately trenchant, the main thesis of this chapter is that *labour is the first algorithm*.

The two main parts of the book endeavour to study machine intelligence in two historical epochs, signalling the parallel development of similar problematics. Part I is concerned with labour as a source of knowledge and with the *automation of mental labour* during the industrial age in the UK. This historical moment is usually studied from the perspective of manual labour, capital accumulation, and fossil energy, and rarely in its cognitive components. Part II, on the other hand,

---

43 Jean-Luc Chabert (ed.), *A History of Algorithms: From the Pebble to the Microchip*, Berlin: Springer, 1999, 1.

Introduction: AI as Division of Labour

analyses the rise of connectionism (the doctrine of artificial neural networks) in the circles of US cybernetics between the 1940s and 1960s. Artificial neural networks emerged from the project of the *automation of visual labour* (commonly termed as pattern recognition), which is something distinct from manual and mental labour. The study of the role of knowledge, mental labour, and science in the nineteenth century is necessary, I contend, to understand the history of automation that prepared the rise of AI in the twentieth century. Under different rubrics, the two parts of the book deal with the same problem: the relation between the forms of technological innovation and social organisation.

As already expounded by historians of science such as Daston and Schaffer, it is easier to find the impetus for modern computation in the workshops of the industrial age than in the volumes of mathematics or natural philosophy of the time. Chapter 2, in this sense, revisits Babbage's pioneering experiments in automated computation – the Difference and the Analytical Engines – focusing on their economic matrix and avoiding the usual machine hagiography. In order to understand the design of these early computers (and their variant of 'machine intelligence'), the chapter explicates two of Babbage's principles of labour analysis. His first analytical principle, the *labour theory of the machine*, states that the design of a machine imitates and replaces the diagram of a previous division of labour. The second, the *principle of labour calculation* (usually called the 'Babbage principle') states that the division of labour into small tasks makes it possible to measure and purchase the exact quantity of labour that is necessary for production. These two principles, combined together, describe the industrial machine not only as a means for augmenting labour but also as an *instrument* (and implicit *metrics*) for measuring it. Babbage applied both principles to the automation of hand calculation: computation emerged, then, not only as the automation of mental labour but also as a metrics for the calculus of its cost.

Beyond the usual 'thermodynamic' interpretations of manual labour, chapter 3 points out that sophisticated notions of mental labour, collective intelligence, and knowledge alienation were already elaborated in the industrial era. It examines the circulation of ideas between the making of nineteenth-century political economy and the Mechanics' Institute movement, between the March of Intellect campaign and the Machinery Question (a debate that animated English society about technological unemployment). The chapter expands, from opposite angles, the previous reflections on Babbage's principles of labour analysis and invention. On the one hand, it shows that, well before the theoreticians of the knowledge society of the twentieth century, a *knowledge theory of labour* was already advanced by Ricardian socialists such as William Thompson and Thomas Hodgskin. On the other hand, it urges a recognition of the influence of industrial machines and instruments on the development of the knowledge of nature, expanding on a *machine theory of science*. The expression 'machine intelligence' ultimately acquires at least four meanings in this discussion: the human knowledge of the machine, the knowledge embodied by the machine's design, the human tasks automated by the machine, and the new knowledge of the world made possible by its use.

Chapter 4 centres on the relation between Babbage and another pillar of the political economy of the industrial age, Karl Marx – a relation which remains under-investigated.[44] The chapter, like every other in this book, explores the imbrication of knowledge into material acts and artefacts, also reading Marx's theories under this lens. In a famous fragment from the *Grundrisse*, Marx predicted that the progressive accumulation of knowledge (or what he called the 'general intellect') into machines would undermine the laws of capitalist accumulation and cause its ultimate crisis. Especially thanks to the interpretation of Italian *operaismo* after 1989, this unorthodox passage (renamed as 'The Fragment on Machines') has had a vast reception among generations of scholars and activists as prophesising the knowledge economy, the dotcom crash, or the rise of AI. The chapter uncovers, after decades of speculation, the origin of the idea of the 'general intellect' – which Marx first encountered in William Thompson's book *An Inquiry into the Principles of the Distribution of Wealth* (1824). The chapter explains, more importantly, why this notion then disappeared in Marx's *Capital*. In Thompson, Marx found the idea of the virtuous accumulation of knowledge but also the argument that once knowledge has been alienated by machines, it becomes hostile to workers. But it was in Babbage that Marx found an

---

44 Exceptions include George Caffentzis, *In Letters of Blood and Fire: Work, Machines, and Value*, Oakland: PM Press, 2013; Amy E. Wendling, *Karl Marx on Technology and Alienation*, Berlin: Springer, 2009. See also Rob Beamish, *Marx, Method, and the Division of Labour*, Urbana: University of Illinois Press, 1992; Bernhard Dotzler, *Diskurs und Medium I: Zur Archäologie der Computerkultur*, Munich: Fink Verlag, 2006.

Introduction: AI as Division of Labour

alternative theory to resolve the ambiguous role that knowledge and science had in the industrial economy. In *Capital*, Marx replaced the utopian expectations around the 'general intellect' with the material figure of the 'general worker' (*Gesamtarbeiter*), which was another name for the extended cooperation of labour. The figure of the general worker, as a sort of super-organism connecting humans and machines, marks in this book the passage to the age of cybernetics and its experiments in self-organisation. As a transition to the second part, chapter 5 briefly summarises the transformation of labour from the industrial to the cybernetic age, clarifying its bifurcation into *abstract energy* and *abstract form* (or *information*).

Part II focuses on connectionism as the main genealogy of current AI systems (avoiding reiterating known literature on cybernetics, information theory, and symbolic AI). Chapter 6 frames the rise of artificial neural networks from a neglected perspective – that is, from the studies on the self-organisation of organisms and machines (which have passed unnoticed even to Boden in her extensive history of AI). Theories of self-organisation are today popular in physics, chemistry, biology, neuroscience, and ecology, but it took cybernetics, rather than a natural science, to trigger the debate on self-organisation in the mid-twentieth century. The chapter illustrates the paradigms of self-organising computation that have contributed, among others, to the consolidation of connectionism – in particular, Warren McCulloch and Walter Pitts's original idea of neural networks (1943–47), John von Neumann's cellular automata (1948), and Rosenblatt's 'perceptron' (1957). The chapter investigates how cybernetic theories of self-organisation also responded to sociotechnical changes. As happened in previous centuries with other variants of mechanistic thinking, cybernetics projected onto brains and nature forms of organisation that were already part of the technical composition of the surrounding society. A key example is the telegraph network, which was used, in the nineteenth century, as an analogy for the nervous system and, in the twentieth century, to formalise neural networks – not to mention the Turing machine itself.

Chapter 7 retraces McCulloch and Pitts's idea of artificial neural networks to the forgotten Gestalt controversy: the debate on whether or not human perception is an act of cognition that can be analytically represented and therefore mechanised. Textbooks on machine learning usually repeat that McCulloch and Pitts were inspired by the neurophysiology of the brain, while overlooking this intellectual confrontation. It was in the aftermath of these debates, in fact, that the expression 'Gestalt perception' gradually morphed, in military and academic publications, into the well-known expression 'pattern recognition'. The Gestalt controversy is a cognitive fossil of unresolved problems whose study can help to understand the form and limits that deep learning has inherited – specifically, the unresolved opposition between perception and cognition, image, and logic, that haunted the technoscience of the twentieth century.

Chapter 8 clarifies the ambivalent role that the neoliberal economist Friedrich Hayek had in consolidating connectionism. In his 1952 book *The Sensory Order*, Hayek proposed a connectionist theory of the mind which was already far more advanced than the definitions of AI that emerged from the 1956 Dartmouth workshop. In this text, as McCulloch and Pitts had also proposed, Hayek speculated about the possibility of a machine fulfilling a similar function to 'the nervous system as an instrument of classification'.[45] Hayek studied the self-organisation of the mind in a similar fashion to the cyberneticians but in order to serve a different agenda: not industrial automation but the autonomy of the market.

Chapter 9 focuses on one of the most important and least studied episodes in the history of AI: Rosenblatt's invention of the 'perceptron' artificial neural network in the 1950s. In spite of its limitations, the perceptron constituted a breakthrough in the history of computation because it automated, for the first time, a technique of statistical analysis; it is remembered, for this reason, as the first algorithm of machine learning.[46] As a technical form, the perceptron claimed to imitate biological neural networks. But as a mathematical form, it expressed a different trick: in order to solve pattern recognition, it represented the pixels of an image as independent coordinates in a multidimensional space. Interestingly, the statistical method of multidimensional

---

45 Friedrich Hayek, *The Sensory Order: An inquiry into the Foundations of Theoretical Psychology*, Chicago: University of Chicago Press, 1952, 55.

46 For the first use of the term 'machine learning' see Arthur Samuel, 'Some Studies in Machine Learning Using the Game of Checkers', *IBM Journal of Research and Development* 44 (1959): 206–26. Also, Turing speculated about 'unorganised machines' that are capable of self-organisation and, in this way, learning: Alan Turing, 'Intelligent Machinery' (1948), in *The Essential Turing*, ed. B. Jack Copeland, Oxford: Oxford University Press, 2004.

projection originated from the fields of psychometrics and eugenics in the late nineteenth century, and was analogous to the technique employed by Charles Spearman for evaluating 'general intelligence' in the controversial practice of the intelligence quotient (IQ) test. This is a further proof of the social genealogy of AI: the first artificial neural network – the perceptron – was born not as the automation of *logical reasoning* but of a statistical method originally used to *measure intelligence* in cognitive tasks and to organise social hierarchies accordingly.

The conclusion considers how the operative principle of AI, in fact, is not just labour automation but also the imposition of social hierarchies of manual and mental labour *through* automation. From the nineteenth century to the twentieth, the 'eye of the master' of industrial capitalism extended to the whole society and imposed new forms of control, also based on statistical measurements of 'intelligence', to discriminate workers into classes of skill. This was, for instance, one of the direct applications of the IQ test according to the US psychologist Lewis Terman, who argued in 1919 that 'the IQ of 75 or below belongs ordinarily in the unskilled labor class, that 75 to 85 is preeminently the range for semi-skilled labor, and that 80 or 85 is ample for success in some kinds of skilled labor'.[47] AI continues this process of encoding social hierarchies and discriminating among the labour force by imposing indirectly a metrics of intelligence. The class, gender, and racial bias that AI systems notoriously amplify should not only be considered a technical flaw, but an intrinsic discriminatory feature of automation in a capitalist context. The impact of AI bias is not limited to social oppression: it also leads to an implicit imposition of labour and knowledge hierarchies that reinforces the polarisation of skilled and unskilled workers in the job market. The replacement of traditional jobs by AI systems should be studied together with the displacement and multiplication of precarious, underpaid, and marginalised jobs across a global economy.[48] AI and ghost

---

47 Lewis Terman, *The Intelligence of School Children: How Children Differ in Ability, the Use of Mental Tests in School Grading, and the Proper Education of Exceptional Children*, Boston: Houghton Mifflin, 1919, 274. Quoted in Stephen Jay Gould, *The Mismeasure of Man*, New York: Norton & Company, 1981, 212.

48 About the deterioration of the job market towards nonsensical occupations, see David Graeber, *Bullshit Jobs: A Theory*, London: Simon & Schuster, 2018. See also Aaron Benanav, *Automation and the Future of Work*, London: Verso, 2020.

work appear to be, in this view, the two sides of the one and same mechanism of labour automation and social psychometrics.

This book proposes the labour theory of automation, in the end, not only as an analytical principle to dismantle the 'master algorithm' of AI monopolies but also as a synthetic principle: as a *practice of social autonomy* for new forms of knowledge making and new cultures of invention.

1
The Material Tools of Algorithmic Thinking

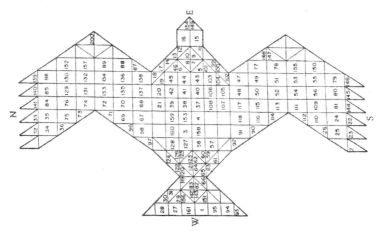

Figure 1.1. Diagram of the Agnicayana fire altar. Frits Staal, 'Greek and Vedic Geometry', *Journal of Indian Philosophy* 27, no. 1 (1999): 111 (image rotated).

The power of our 'mental' tools is amplified by the power of our 'metal' tools.

Jeannette Wing, 'Computational Thinking', 2008[1]

---

1 Jeannette M. Wing, 'Computational Thinking and Thinking about Computing', *Philosophical Transactions of the Royal Society A: Mathematical, Physical, and Engineering Sciences* 366, no. 1881 (2008): 3718.

When using a material tool, more can always be learned than the knowledge invested in its invention.
  Peter Damerow and Wolfgang Lefèvre, 'Tools of Science', 1981[2]

Rules became mechanical before they could actually be executed by machines.
  Lorraine Daston, 'Algorithms before Computers', 2017[3]

Recomposing a dismembered god

In a myth of cosmogenesis of the Vedas, it is narrated that the supreme god Prajapati is shattered into pieces in the act of creating the universe. In the aftermath of creation, counter-intuitive to Western narratives of mastery and principles of non-contradiction, the creator's body is found unstrung, dismembered. This ancient myth is still re-enacted today, in India, in the Agnicayana ritual, in which Hindu devotees symbolically recompose the fragmented body of the god by building the fire altar Syenaciti (see fig. 1.1). The Syenaciti altar is laid down by aligning a thousand bricks of precise shape and size according to an elaborate geometric plan that draws the profile of a falcon. Workers compose five layers of 200 bricks each while reciting dedicated mantra and following step-by-step instructions. Solving a riddle that is the key to the ritual, each layer must maintain the same area and shape but a different configuration.[4] Finally, the falcon altar must face east, in prelude to a symbolic flight of the reconstructed god towards the rising sun – a unique example of divine reincarnation by geometric means.

Agnicayana is meticulously described in the appendices to the Vedas dedicated to geometry, the Shulba Sutras, which were composed around

---

  2 Peter Damerow and Wolfgang Lefèvre, 'Tools of Science', in Peter Damerow, *Abstraction and Representation: Essays on the Cultural Evolution of Thinking*, Berlin: Springer, 2013, 401.

  3 Lorraine Daston, 'Algorithms before Computers: Patterns, Recipes, and Rules', Katz Distinguished Lecture in the Humanities, Simpson Center for the Humanities, University of Washington, 19 April 2017. See also: Lorraine Daston, *Rules: A Short History of What We Live By*, Princeton, NJ: Princeton University Press, 2022.

  4 K. Ramasubramanian, 'Glimpses of the History of Mathematics in India', in *Mathematics Education in India: Status and Outlook*, ed. R. Ramanujam and K. Subramaniam, Mumbai: Homi Bhaba Centre for Science Education (TIFR), 2012.

800 BCE in India, yet recording a much-older oral tradition.[5] They narrate that the *rishi* (vital spirits) created seven square-shaped *purusha* (cosmic beings) which together composed a single body, and it is from this simple configuration that the complex body of Prajapati evolved.[6] The Shulba Sutras teach the construction of other altars of specific geometric forms to secure the auspices of gods. They suggest, for instance, that 'those who wish to destroy existing and future enemies should construct a fire-altar in the form of a rhombus'.[7] Beyond religious symbolism, the Agnicayana ritual and the Shulba Sutras in general had, in fact, the function of transmitting useful techniques for the society of the time, such as how to plan a construction and to enlarge existing buildings while maintaining their original proportions.[8] Agnicayana exemplifies the originary social materiality of mathematical knowledge but also the hierarchies of manual and mental labour typical of a caste system. In the construction of the altar, the workers are driven by rules which are traditionally possessed and transmitted only by a specific group of masters. Aside from geometric exercises, rituals such as Agnicayana taught a kind of procedural knowledge which is not just abstract but based on continuous 'mechanical' drill, pointing once again to the role of religion as a motivation for exactness and, at the same time, to spiritual exercises as a way of disciplining labour.[9]

---

5 Agnicayana has been reported by the Dutch indologist Frits Staal in two volumes and a documentary about an expedition in Kerala, India, in 1975: Frits Staal, *Agni: The Vedic Ritual of the Fire Altar*, two vols., Berkeley: Asian Humanities Press, 1983. Staal argued that abstract cultural forms emerge as nonconscious, and that language, numerals, and geometry are first collective practices. See Frits Staal, *Rules without Meaning: Ritual, Mantras, and the Human Sciences*, New York: Peter Lang, 1989, 71.

6 Paolo Zellini, *La matematica degli dèi e gli algoritmi degli uomini*, Milano: Adelphi, 2016, 41. Translated as *The Mathematics of the Gods and the Algorithms of Men*, London: Penguin, 2020.

7 Kim Plofker, 'Mathematics in India', in *The Mathematics of Egypt, Mesopotamia, China, India, and Islam*, ed. Victor Katz, Princeton, NJ: Princeton University Press, 2007.

8 For a study of knowledge and technology transfer in antiquity, see Jürgen Renn (ed.), *The Globalization of Knowledge in History*, Princeton, NJ: Princeton University Press, 2020.

9 The division of labour of Agnicayana is also reminiscent of the elaborate *chaîne opératoire* (operational chain) that the French anthropologist André Leroi-Gourhan has identified in many ancestral practices of tool-making which originally are not hierarchical but spontaneous and cooperative. See Frederic Sellet, 'Chaîne opératoire: The Concept and its Applications', *Lithic Technology* 18, nos. 1–2 (1993): 106–12.

Agnicayana is a unique artefact in the history of human civilisation: it is the most ancient documented ritual of humankind that is still practised today – although, due to its complexity, it is performed only a few times in a century.[10] Across all this time, it has transmitted and preserved sophisticated paradigms of knowledge, and because of its combinatorial mechanism, it can be defined as a primordial example of algorithmic culture. But how can one possibly interpret a ritual as ancient as Agnicayana as *algorithmic*? One of the most common definitions of algorithm in computer science is the following: a finite procedure of step-by-step instructions to turn an input into an output, independently of the data, and making the best use of the given resources.[11] The recursive mantras which guide workers in the construction site of the fire altar may indeed resemble the rules of a computer program: independently of the context, the Agnicayana algorithm organises a precise distribution of bricks which results every time in the construction of the Syenaciti. Historians have found that Indian mathematics has been predominantly algorithmic since ancient times, meaning that the solution to a problem was proposed via a step-by-step procedure rather than a logical demonstration.[12]

Similarly, the Italian mathematician Paolo Zellini has argued that the Agnicayana ritual evidences a more sophisticated technique than simple obedience to a rigid rule, namely the *heuristic technique of incremental approximation*. It is known that Vedic mathematics, before other civilisations, was familiar with infinitely large and infinitesimally small numbers: ancient sutras already multiplied the positional numerals of the Hindu system to large scales to signify the vast dimensions of the universe (a speculative exercise that would be impracticable with the additive systems of Sumerian, Greek, and Roman numerals, for instance). Vedic mathematics was also familiar with irrational numbers, such as the square root, which in many cases (such as $\sqrt{2}$) can only be

---

10 The last were in 1955, 1975 (the ceremony documented by Frits Staal), and 2011.

11 See also: 'An algorithm is a finite sequence of rules to apply in a determined order to a finite set of data to arrive, in a finite number of steps, at a certain result, independently of the data'. Jean-Luc Chabert (ed.), *A History of Algorithms: From the Pebble to the Microchip*, Berlin: Springer, 1999, 2. My translation: the French original provides a more precise definition as the English edition removes 'independently of the data'. Jean-Luc Chabert, *Histoire d'algorithmes: Du caillou à la puce*, Paris: Belin, 1994, 6.

12 M. S. Sriram, 'Algorithms in Indian Mathematics', in *Contributions to the History of Indian Mathematics*, Gurgaon: Hindustan Book Agency, 2005, 153–82.

calculated by approximation. The mantras of the Shulba Sutras intone the most ancient (and pedantic) explications of computational procedures (like to the so-called Babylonian algorithm) to approximate square root results. Procedures of approximation may appear cumbersome, weak, and imprecise compared to the exactitude of our mathematical functions and geometric theorems, but their role within the history of mathematics and technology is more important than is commonly thought. In his history of the techniques of incremental growth (which include the ancient method of gnomon, among others), Zellini has argued that the ancient Hindu techniques of incremental approximation are equivalent to the modern algorithms of Leibniz and Newton's calculus, and even to the error-correction techniques that are found at the core of artificial neural networks and machine learning, which constitute the current paradigm of AI (see chapter 9).[13]

To some, it may appear an act of misappropriation to read ancient cultures through the paradigm of the latest technologies from Silicon Valley or to study the mathematical component of religious rituals in an age of rampant nationalism. However, to claim that abstract techniques of knowledge and artificial metalanguages belong uniquely to the modern industrial West is not only historically inaccurate but an act of implicit *epistemic colonialism* towards the cultures of other places and other times.[14] Thanks to the contribution of ethnomathematics, decolonial studies, and the history of science and technology, alternative forms of computation are now recognised and investigated outside the Global North hegemony and its regime of knowledge extractivism. Because of their role in computer programming, algorithms are usually perceived as the application of complex sets of rules in the abstract; on the contrary, I argue here that algorithms, even the complex ones of AI and machine

---

13 Zellini, *La matematica degli dèi*, 51. For a contested but influential history of calculus, see Hermann Cohen, *Das Prinzip der Infinitesimal-Methode und seine Geschichte: Ein Kapitel zur Grundlegung der Erkenntniskritik* (1883).
14 The historian of mathematics Senthil Babu remarks: 'The history of mathematics in India has thus far primarily been an engagement with a corpus of texts recorded in Sanskrit . . . Indology recognized and canonized only the dignified Sanskritic tradition. The knowledge of many practitioners of mathematics was rendered invisible.' Senthil Babu, *Mathematics and Society: Numbers and Measures in Early Modern South India*, Oxford: Oxford University Press, 2022, 2–5. See also Senthil Babu, 'Indigenous Traditions and the Colonial Encounter: A Historical Perspective on Mathematics Education in India', in Ramanujam and Subramaniam, *Mathematics Education in India*.

learning, have their genesis in social and material activities. Algorithmic thinking and algorithmic practices, broadly understood as rule-based problem-solving, have been part of all cultures and civilisations.

Along these lines of inquiry, this chapter sketches a provisional history of algorithms, broadly examining in turn (1) *social algorithms*, that is, procedures that were embodied in rituals and practices, often transmitted orally and not formalised into symbolic language; (2) *formal algorithms*, that is, mathematical procedures to help calculation and administrative operations as they are found, for instance, in Europe since the Middle Ages and before that in India; and (3) *automated algorithms*, that is, the implementation of formal algorithms in machines and electronic computers starting with the industrial age in the West.

Archaeology of the algorithm

The idea of investigating 'algorithms before computers' first came, unsurprisingly, from the field of computer science. In the late 1960s, the US mathematician Donald Knuth authored the influential book *The Art of Computer Programming* and gave important contributions to excavating the deep time of mathematical techniques in essays such as 'Ancient Babylonian Algorithms'. In those years, Knuth's mission was to systematise the field of computer science and to make it into a respectable academic discipline. The evidence of ancient algorithms was mobilised to stress that computer science was not about obscure electronic apparatuses but part of a long tradition of cultural techniques of symbolic manipulation. In this case, however, the archaeology of the algorithm was pursued not to demonstrate universalistic principles of thinking or the emancipatory potential of learning across the history of civilisation, but for the specific interests of the new classes of computer programmers and manufacturers:

> One of the ways to help make computer science respectable is to show that it is deeply rooted in history, not just a short-lived phenomenon. Therefore it is natural to turn to the earliest surviving documents which deal with computation, and to study how people approached the subject nearly 4000 years ago. Archeological expeditions in the Middle East have unearthed a large number of clay tablets which contain

mathematical calculations, and we shall see that these tablets give many interesting clues about the life of early 'computer scientists'.[15]

Knuth observed that mathematical formulas that today would be defined as algebraic or analytical were already described by the Babylonians through step-by-step procedures, namely algorithms. These procedures were, of course, formulated in the words of the common language and not yet in the symbolic metalanguage of mathematics. Knuth's research confirms the hypothesis that procedure-based methods (what he called a 'machine language') predated the consolidation of mathematics as a metalanguage of symbolic representations:

> The Babylonian mathematicians were not limited simply to the processes of addition, subtraction, multiplication, and division; they were adept at solving many types of algebraic equations. But they did not have an algebraic notation that is quite as transparent as ours; they represented each formula by a step-by-step list of rules for its evaluation, i.e. by an algorithm for computing that formula. In effect, they worked with a 'machine language' representation of formulas instead of a symbolic language.[16]

Knuth intended to liberate the algorithm from the age of computer science and engineering in order to make it, retroactively, a broad subject for the history of culture. This happened in the 1960s, when computer science was still struggling, as the historian Nathan Ensmenger has highlighted, to achieve the status of proper discipline in the United States. This qualification became possible by establishing as its central concept the algorithm, rather than information as happened in Europe (see the German *Informatik*, the French *informatique*, and the Italian *informatica* as names for computer science).[17] This canonization of the algorithm is particularly significant for the historians of science and technology because it proceeded from within its original professional milieu: the operators of computing machines, a new

---

15 Donald E. Knuth, 'Ancient Babylonian Algorithms', *Communications of the ACM* 15, no. 7 (1972): 671.

16 Ibid., 672.

17 Nathan Ensmenger, *The Computer Boys Take Over: Computers, Programmers, and the Politics of Technical Expertise*, Cambridge, MA: MIT Press, 2010, 131.

generation of mental workers, were up to write their own history of technology – and obviously they did it according to the logical form their work embodied.

The reconstruction of the prehistory of the algorithm (one may say its 'archaeology') has also been a resurgent concern in mathematics. Notably, the French mathematician Jean-Luc Chabert has contributed an exemplary synthesis that also ventures beyond the disciplinary borders of computer science:

> Algorithms have been around since the beginning of time and existed well before a special word had been coined to describe them. Algorithms are simply a set of step by step instructions, to be carried out quite mechanically, so as to achieve some desired result ... Algorithms are not confined to mathematics ... The Babylonians used them for deciding points of law, Latin teachers used them to get the grammar right, and they have been used in all cultures for predicting the future, for deciding medical treatment, or for preparing food ... We therefore speak of recipes, rules, techniques, processes, procedures, methods, etc., using the same word to apply to different situations. The Chinese, for example, use the word *shu* (meaning rule, process or stratagem) both for mathematics and in martial arts ... In the end, the term algorithm has come to mean any process of systematic calculation, that is a process that could be carried out automatically. Today, principally because of the influence of computing, the idea of finiteness has entered into the meaning of algorithm as an essential element, distinguishing it from vaguer notions such as process, method or technique.[18]

Also in this reading, the algorithm does not appear to be the most recent technological abstraction but a very ancient technique – one that predates many tools and machines that the human mind has designed. These efforts of historicisation invite a reconsideration of the algorithm, then, as a fundamental *cultural technique* of humankind, which gradually emerged from collective practices and rituals temporally very close to the constituent and primordial traits of all civilisations. The algorithm should be added, in summary, to the list of techniques that the historian of culture Thomas Macho compiled in an often-quoted passage:

---

18 Chabert, *A History of Algorithms*, 1.

Cultural techniques – such as writing, reading, painting, counting, making music – are always older than the concepts that are generated from them. People wrote long before they conceptualized writing or alphabets; millennia passed before pictures and statues gave rise to the concept of the image; and until today, people sing or make music without knowing anything about tones or musical notation systems. Counting, too, is older than the notion of numbers. To be sure, most cultures counted or performed certain mathematical operations; but they did not necessarily derive from this a concept of number.[19]

The research on cultural techniques (in German, *Kulturtechniken*, which can also be translated as 'techniques of civilisation') has stressed the role of material practices in the making of all the symbolic forms of civilisations. This open-minded and pluralistic view, however, often neglects to study the causes of this evolution towards abstraction, resulting in a culturalist interpretation of what are more profound phenomena. Chabert, in his history of algorithms, for example, relates the rise of techniques of calculation to economic needs: 'The basic arithmetic operations of the elementary school, multiplying and dividing, appear to have derived from extremely early economic needs, certainly earlier than the emergence of civilisation using writing.'[20] Although it is always difficult to generalise historical findings about the remote past, economic problems – such as conditions of lack or surplus of resources – appear to be at the origins of counting and mathematical techniques.[21] It is worth remembering, with no intention of reviving ancestral famines, that the word 'number' comes from the Latin *numerus*, or 'portion of food'.

Well before the institution of mathematical and geometric disciplines, ancient civilisations were already large 'machines' of social segmentation, marking human bodies and territories with abstractions that would remain operative for millennia. It is known and repeated that one of the first recorded censuses of the population, organised by the

---

19 Thomas Macho, 'Zeit und Zahl: Kalender- und Zeitrechnung als Kulturtechniken', in *Bild – Schrift – Zahl*, ed. Sybille Krämer and Horst Bredekamp, Munich: Wilhelm Fink, 179. Quoted in translation in Geoffrey Winthrop-Young, 'Cultural Techniques: Preliminary Remarks', *Theory, Culture, and Society* 30, no. 6 (2013): 8.

20 Chabert, *A History of Algorithms*, 7.

21 For an alternative analysis of primitive economies, see Marshall Sahlins, *Stone Age Economics*, Chicago: Aldine-Atherton, 1972.

Babylonians, took place around 3800 BCE, but history records that these 'cultural techniques' were also inhuman and ruthless. Drawing on historian Lewis Mumford and his account of ancient societies as 'megamachines', Gilles Deleuze and Félix Guattari enumerated other techniques of abstraction than number on which social order was based. They argue that in ancient civilisation, the power to control 'productive forces . . . resides in these operations: tattooing, excising, incising, carving, scarifying, mutilating, encircling, and initiating'.[22] Numbers and counting tools were components of these *primitive abstract machines* that forged human civilisations through territorialisation and segmentation. Numbers, as much as abstract rules and heuristic practices, were key tools in the administration of ancient societies, but they were not invented from nothing: they materially emerged as a form of power through labour and rituals, through discipline and drill.

This intrinsic relation between mathematical abstractions and material life was not overlooked even by a neo-Kantian philosopher like Ernst Cassirer, who has exercised quite an influence on cultural studies in German-speaking countries. According to Cassirer, the 'symbolic form' of number emerged from the relation of the human body with its environment and the contingent use of the body as the first medium of calculation: 'It is through material enumerable things, however sensuous, concrete and limited its first representation of these things may be, that language develops the new form and the new logical force that are contained in number.'[23] Analysing the perception of space and time, Cassirer traced the origin of numerical abstractions to the rhythmical activities of work. Following Karl Bücher's seminal book *Arbeit und Rhythmus* (1896) and other anthropological studies, Cassirer remarked that the symbolic form of number grew out of the custom of work songs – that is, singing to sustain the rhythm of work:

> Attempts have been made to trace the beginnings of poetry back to those first primitive *work songs* in which for the first time the rhythm felt by man in his own physical movements was, as it were,

---

22 Gilles Deleuze and Felix Guattari, *Anti-Oedipus: Capitalism and Schizophrenia*, New York: Viking, 1977, 145.

23 Ernst Cassirer, *The Philosophy of Symbolic Forms*, vol. 1, New Haven, CT: Yale University Press, 1965, 228. Based on Max Wertheimer, *Über das Denken der Naturvölker*, Leipzig: Barth, 1910.

objectified … Every form of physical labor, particularly when performed by a group, occasions a specific coordination of movements, which leads in turn to a rhythmic organization and punctuation of work phases … Grinding and rubbing, pushing and pulling, pressing and trampling: each is distinguished by a rhythm and tone quality of its own. In all the vast variety of work songs, in the songs of spinners and weavers, threshers and oarsmen, millers and bakers, etc., we can still hear with a certain immediacy how a specific rhythmic sense, determined by the character of the task, can only subsist and enter into the work if it is at the same time objectified in sound … In any case, language could acquire consciousness of the pure forms of time and number only through association with certain contents, certain fundamental rhythmic experiences, in which the two forms seem to be given in immediate concretion and fusion.[24]

This study can be taken as a rejoinder to the Platonic numerology that is central to the history of music: before numbers were used to measure the proportions of rhythm, the rhythm of work contributed to the invention of numbers. At the end, these findings cast a different light on the history of mathematics, so much that one could suspect, at this point, that algorithmic practices are even older than the concept of number itself.

Tools for the construction of mathematical ideas

Numbers are often considered as something given, originary and elemental, not composed of anything else and not resulting from any prior conceptual fabrication. Numbers appears to be self-explanatory, eternal, and not constructed. Such a Platonic and intuitionist view of the concept of number has been criticised by historians of mathematics, who are particularly concerned with explaining how techniques of numeration arose and evolved. Archaeologists, especially, are inclined to suggest that the institution of number cannot be an a priori category, as human activity with materials and symbolic tools testifies to its gradual evolution: counting appeared to have emerged, as already mentioned,

---

24 Ibid., 240.

from the need to calculate and solve practical problems, such as the equal distribution of land and natural resources in the population.

Among the *archaeologists of abstraction* we encounter the German historian of science Peter Damerow, who extensively studied, among other artefacts, ancient Babylonian clay tablets that were used as counting tools. Damerow came to the conclusion that the idea of number is not a form of a priori knowledge but 'subject to historical development'.

> Reflections on numbers and their properties led already in antiquity to the belief that propositions concerning numbers have a special status, since their truth is dependent neither on empirical experience nor on historical circumstances. In a historical tradition extending from the Pythagorean through the Platonic tradition of Antiquity, Late Antiquity and the Middle Ages, further through the rationalism and the critical idealism of Kantian and neo-Kantian philosophy to the logical positivism and constructivism of the present, this belief has been considered proof that there are objects of which we can gain knowledge *a priori*. Like a recurring leitmotif, the conviction that numbers are by nature ahistorical and universal is woven through the history of philosophy. A variety of reasons have been proposed to explain this puzzling phenomenon. The historian, on the other hand, is confronted with the fact that numerical techniques and arithmetic insights have a history that is, at least on its surface, in no way different from other achievements of our culture. In view of the variety of historically documented arithmetical techniques, it is scarcely possible to dismiss the assumption that the concept of number – in the same way as most structures of human cognition – is subject to historical development, which in the course of history exposes it to substantial change.[25]

Engaging with the findings of archaeology, moreover, Damerow realised that 'the emergence of numbers appears as the result of manifold

---

25 Peter Damerow, 'The Material Culture of Calculation: A Theoretical Framework for a Historical Epistemology of the Concept of Number', in *Mathematisation and Demathematisation: Social, Philosophical, and Educational Ramifications*, ed. Uwe Gellert and Eva Jablonka, Leiden: Brill, 2008, 19.

learning processes'.[26] Learning became a central notion in Damerow's research, through which he explained the making of human civilisation and its evolution. For Damerow, learning is a process of interaction of humankind with nature and the world, mediated by labour, tools, and language in a continuous process of abstraction. Learning, however, is not a process of abstraction for the sake of abstraction but a collective means of emancipation and empowerment. How does this social process of learning take place?

Damerow argued that learning is based on the construction of 'mental models' that fundamentally represent and internalise external actions.[27] On top of these internalised mental models, further levels of abstraction can be built in a progressive scaffolding of 'meta-cognitive constructs'.[28] This continuous scaffolding of abstractions is a form of the emancipation of reason, but it happens that some levels are eventually perceived as metaphysical and separated from others. According to Damerow, the higher levels of the cognitive scaffolding create the illusion of dematerialised abstractions and a priori categories such as the concept of number. However, what is decisive in this theory is not simply the explanation of the a priori illusion but rather how 'mental operations . . . reflect actions on real objects' and, vice versa, how tools help constructing mental models:

> Logico-mathematical concepts are abstracted not directly from the objects of cognition, but from the coordination of the actions that they are applied to and by which they are somehow transformed. According to this assumption the emergence of mental operations of logico-mathematical thought is based on the internalisation of systems of real actions. The internalised actions are the starting-point for meta-cognitive constructions, through which they become elements of systems of reversible mental transformations which, following Piaget's terminology, we will call here 'operations'. Meta-cognitive constructs such as the concept of number that are generated by reflective abstractions can thus be understood as internally represented invariables of mental operations which reflect actions on real

---

26 Ibid., 20.
27 Ibid.
28 Ibid., 22.

objects. This explains the puzzling *a priori* nature of constructions such as the number concept.[29]

To explain the formation of the concept of number throughout history, Damerow suggested a scaffolding of semiotic and cognitive models that progressively unfolded from *practices of counting* (which are heuristic and non-formalised, such as reckoning with fingers), to *systems of numeration* (which represent quantities in a matrix of symbols), to *techniques of computation* (which express algorithms or procedure to solve problems by manipulating symbols), and eventually to *number theory* (namely arithmetic as a formal discipline). This process is not linear but unfolds, according to Damerow, through an alternate movement of *representation* (the use of objects and signs as symbols of other objects, signs, and ideas) and *abstraction* (problem-solving).

Applying the idea of *reflective abstraction* that combines both Hegel's dialectical logic and Jean Piaget's genetic epistemology, Damerow sketched progressive stages of symbolic representation (see fig 1.2), in which the passage from one order of representation to the following occurs via the solution of a problem. According to Damerow,

> *First order representations* are representations of real objects by symbols or models which permit the performance of essentially the same actions or operations with these symbols as can be performed with the real objects themselves ... *Second and higher order representations* are representations of mental objects by symbols and symbol transformation rules which correspond to mental operations belonging to the cognitive structures constituting the mental objects.[30]

The concept of number developed, then, through cycles of symbolic representation and abstraction. First, processes of quantification and comparison that were based on equivalence without involving counting. Counting then emerged as a *context-dependent activity* that utilised aids such as fingers, stones, and so forth. Thereafter, these counting devices

---

29 Ibid.
30 Peter Damerow, 'Abstraction and Representation', in *Abstraction and Representation: Essays on the Cultural Evolution of Thinking*, Berlin: Springer, 2013, 373.

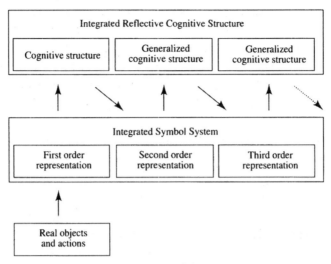

Figure 1.2. The reflective structure of abstraction. Peter Damerow, *Abstraction and Representation* Berlin: Springer, 2013, 379.

were replaced by *context-dependent symbols* (such as the signs on the bullae for trading in ancient Mesopotamia). Subsequently, *context-free symbols* were introduced, namely numbers in the modern sense. Finally, *arithmetic* emerged as a discipline to describe numbers and operations with natural language words, eventually to be replaced by new symbols themselves.[31]

To understand if such analysis can also be applied to the algorithmic form as a practice of problem-solving, it is necessary, at this point, to clarify Damerow's idea of abstraction. Following Hegel and Piaget, Damerow understood abstraction as a process in which materiality and reflection, that is tools and cognition, are mutually imbricated and mutually evolve:

> The concept [of reflection] was introduced [by Hegel] in the *Jenaer Realphilosophie* to distinguish labor as 'reflective activity' from activity as 'pure mediation', as the mere satisfaction of a desire by means of the destruction of its object. What distinguishes labor as 'reflective activity' from activity as 'pure mediation' is the endurance of its material means, of its tools, in which activity as the unity of ideal purpose

---

31 Damerow, 'Material Culture', 34–47.

and material object has materially objectified itself . . . The unity of the sensory given and mediating activity constructed in Hegel's logic, the mediated immediacy as the result of reflection, not only constitutes a hypothetico-theoretical construct, but this unity is actually created in the material means of the objective activity in a myriad of forms.[32]

For Damerow abstraction is not about isolating the most prominent features of a given structure but about producing new knowledge in relation to a problem to solve: abstraction is not just an 'elegant solution' to a problem but 'an activity directed towards some end or goal', which includes the contingent understanding of the environment:[33]

> It is a common view that abstraction means refraining from using the information available on a given real object, and instead isolating certain properties and dealing with these independently. But this concept of abstraction reveals itself as unsatisfactory if it is used to conceptualize the development of mathematical thinking. Abstraction in this sense does not explain that outcome of new knowledge which obviously does result from mathematical thinking. Furthermore, this concept of abstraction makes it impossible or at least difficult to understand why certain abstractions turn out to be very useful but the huge mass which might be produced by arbitrarily isolating properties of mathematical objects would only result in nonsense . . . To understand abstraction essentially means understanding *what* has to be abstracted rather than merely knowing *how* it has to take place. To understand the abstraction leading to an elegant solution of the problem means understanding how the solution can really be found.[34]

Abstraction always operates within given material constraints and through them: symbols, tools, techniques, and technologies are conceived and realised in relation to limited resources of matter, energy, space, time, and so on. The reality which abstraction is struggling with

---

32 Damerow, 'Action and Cognition in Piaget's Genetic Epistemology and in Hegel's Logic', in Damerow, *Abstraction and Representation*, 8.
33 Damerow, *Abstraction and Representation*, 372.
34 Ibid., 371.

is not the idealised space of Platonic ideas but the actual living world, made of force fields and conflicts. In this sense, abstraction is also part of the larger social antagonism.

Importantly, material constraints give an impetus to expand the reach of abstraction beyond its original field. Together with his colleague Wolfgang Lefèvre, Damerow extended the historical epistemology of mathematics to the relation of science in general with tools and instruments. Their understanding of tools is, at one and the same time, contingent and speculative – in short, dialectical. Tools are not just means to an end but means that exceed the purpose of their initial design:

> Tools determine whether goals anticipated mentally can be realised. In that sense tools are never just what they actually are. Rather, they represent the potential of realizing intellectually anticipated goals, which is to say, they represent ideas as real possibilities. Their application mediates between possibility and reality. The use of tools primarily serves the purpose for which they were produced. But tools are more general than particular purposes, and the accumulated experience acquired in the course of their use leads to knowledge about possibilities capable of being realized and about relationships between goals and means under various conditions of realization. Thus, the primary form in which knowledge about natural and social relationships arising from the labor process is represented is the form of rules for the appropriate use of tools.[35]

In this understanding, the speculative process starts with labour that invents tools and technologies which, subsequently, project new ontological dimensions and scientific fields (a canonical example is the invention of the steam engine that engendered the discipline of thermodynamics, rather than the other way around; see chapter 3). Damerow and Lefèvre advance a political epistemology that acknowledges the constraints of historical forces, namely the control of resources and population, economic production and capital accumulation, the rise of wars and social conflicts, and, because of all of this, the development of new tools, techniques, technologies, and eventually science. They acknowledge all these forces within the category of labour, through

---

35 Damerow and Lefèvre, 'Tools of Science', 395.

which humans transform nature and produce new knowledge about it.[36] Science in general, as much as the concept of number in particular, is a projection of the use of material tools:

> The development of science depends on the development of its material tools... The key of understanding the growth of scientific knowledge consists in the fact that the knowledge to be gained by using a new tool exceeds the cognitive preconditions of its invention. The reason for this is due to the fact that the tools of science like tools in general are material tools: When using a material tool, more can always be learned than the knowledge invested in its invention.[37]

Along this historical overview of material abstractions, one can easily imagine also the concept of algorithm emerging as the result of a dialectical process of reflection with objects and tools. The method of the algorithm – the resolution of a problem by step-by-step instructions – is an abstraction that like many others emerged from the troubles of this world.[38]

From counting tools to algorithms for calculation

The English term 'algorithm' is circa eight centuries old. It derives from the medieval Latin term *algorismus*, which referred to the procedures for executing the basic mathematical operations with the Hindu–Arabic numerals. In the Europe of the Middle Ages, thanks to the trading routes with the Arab world, the limited system of Roman additive numerals came gradually to be replaced by the more versatile Hindu–Arabic positional system, which was more practical for complex operations on large numbers and has since become the planetary standard. The Latin term

---

36 '[The] basic structures of logico-mathematical thought are ... developed by the individual growing up in confrontation with culture-specific challenges and constraints under which the systems of action have to be internalised.' Damerow, 'Material Culture', 22.
37 Damerow and Lefèvre, 'Tools of Science', 400–1.
38 For a discourse analysis of the algorithm concept that does not consider the economic matrix, see Yu Mingyi, 'The Algorithm Concept, 1684–1958', *Critical Inquiry* 47, no. 3 (2021): 592–609.

algorismus is found, for instance, in the 1240 poem 'Carmen de Algorismo' by Alexandre de Villedieu – a manual of calculation techniques that was composed in rhyming verse as an aid to memorise such procedures. A book printed in Venice in 1501 and attributed to the thirteenth-century monk Johannes de Sacrobosco bears the title *Algorismus Domini* and explicates hand calculation using Hindu numerals also with diagrams.

Only recently has it been established that *algorismus* is a Latinisation of the name of the Persian scholar Muhammad ibn Musa al-Khwarizmi, head librarian at the House of Wisdom in Baghdad, who authored a book on calculation with Hindu numerals around 825 CE. Al-Khwarizmi's original in Arabic has been lost, but in the twelfth century at least four Latin translations circulated under different titles: a manuscript at Cambridge University Library bears the incipit 'DIXIT algorizmi' (meaning: 'so spoke Al-Khwarizmi'), while another was given in 1857 the title *Algoritmi de numero Indorum* by Italian mathematician Baldassarre Boncompagni.[39] It is through various transliterations in Romance languages, such as French and Spanish, that the English term 'algorithm' has reached contemporary mathematics and computer science. Al-Khwarizmi's book helped to introduce the positional Hindu numerals to the West, yet merchants, such as the Italian mathematician Fibonacci, who travelled frequently across the Mediterranean, probably learned the system more through commercial exchanges and practice than through books.

In terms of mathematical conventions, the adoption of the term 'algorithm' in the West marked the shift from the additive to the positional system of numeration. This shift was both technical and economic, as it was related to the acceleration of commercial exchanges across Europe and the Mediterranean that demanded a better system of accounting. The decimal positional system made it possible to write numbers more concisely and sped-up calculations. In Italy, Florentine and Venetian merchants were the first to adopt the Hindu numerals, favoured for their greater versatility in commercial transactions and in handling capital's increasingly large figures. A drawing from the 1503

---

39 John N. Crossley and Alan S. Henry, 'Thus Spake al-Khwârizmî: A Translation of the Text of Cambridge University Library Ms. Ii. vi. 5', *Historia Mathematica* 17, no. 2 (1990): 103–31.

Figure 1.3. Allegory of Arithmetic. Gregor Reisch, *Margarita Philosophica*, 1503.

book *Margarita philosophica*, edited by the German monk and polymath Gregor Reisch, shows the dispute between abacists (who were still using the Roman system and the abacus) and the new algorists (who adopted the Hindu system and its algorithms to make calculations on paper with a stylus). The allegory of *Arithmetica* supervises the dispute, clearly deciding in favour of the algorist, her cloth covered with the new numerals (see fig. 1.3). Later, the term 'algorithm' was adopted by the scholars of European high culture, such as Leibniz, who used it to define his method of differential calculus.[40] 'Algorithm' was broadly defined by D'Alembert's *Encyclopédie* as an

> Arab term, used by several authors, and particularly by the Spanish to mean the practice of algebra. It is also sometimes taken to mean arithmetic by digits ... The same word is taken to mean, in general, the method and notation of all types of calculation. In this sense, we say the algorithm of the integral calculus, the algorithm of the exponential calculus, the algorithm of sines, etc.[41]

The techniques and tricks for doing calculations by hand that are still taught in school to this day are a set of algorithms for the manipulation of numerical signs. They possess a recursive structure that can handle infinite and approximate digits, as occurs in the simple division of the prime numbers: $2/3 = 0.666666666 \ldots$ The simple continuous form of this fraction shows that even rational numbers cannot be calculated and expressed without the help of an algorithm. More precisely, even the way in which numbers are written in a system of numeration constitutes an algorithm – in this case, an algorithm to represent simple quantities. For instance, when we write the number 101 in Hindu numerals, this simple sign should be translated as:

> Consider a linear sequence of positions to be occupied by symbols of quantity running from right to left. Each position represents incrementally a power of ten and can be filled by one of the ten units: 0, 1, 2, 3, 4, 5, 6, 7, 8, or 9. The first position represents ten to the power of zero (that is, normal units), the second position ten to the power of one (ten), the third position ten to the power of two (hundred), and so on. The value of a number represented in this way is given by the addition of each unit after being multiplied to the power of ten, represented by the occupied position. When numbers are expressed in this way, the scale of the powers of ten is not explicitly stated but remains implicit.

---

40 See also Sybille Krämer, 'Zur Begründung des Infinitesimalkalküls durch Leibniz', *Philosophia Naturalis* 28, no. 2 (1991): 117–46; Peter Damerow and Wolfgang Lefèvre, 'Wissenssysteme im geschichtlichen Wandel', in *Enzyklopädie der Psychologie. Themenbereich C: Theorie und Forschung*, ed. F. Klix and H. Spada, Serie II: Kognition, Band 6: Wissen, Göttingen: Hogrefe, 1998, 77–113.
41 Quoted in Chabert, *A History of Algorithms*, 2.

The number 101, therefore, is equal to: (1 × hundred) + (0 × ten) + (1 × one). This verbose explanation of the decimal system in natural language can be easily adapted to represent the binary system by simply switching the power of ten to the one of two – that is, by changing one rule in the general procedure of numeration. In the binary system, the number 101 comes to signify a different quantity:

> Consider a linear sequence of positions to be occupied by symbols of quantity running from right to left. Each position represents incrementally a power of two and can be filled by one of the two units: 0 or 1. The first position represents two to the power of zero (that is, normal units), the second position two to the power of one (two), the third position two to the power of two (four), and so on. The value of a number represented in this way is given by the addition of each unit after being multiplied to the power of two, represented by the occupied position. When numbers are expressed in this way, the scale of the powers of two is not explicitly stated but remains implicit.

In this case, the number 101 is equal to: (1 × four) + (0 × two) + (1 × one), that is, 5 in decimal notation. In both of these verbose paraphrases, the words of natural language are not used to explain but to *encode* rules for the construction of numbers with a procedure of step-by-step instructions. These paraphrases make visible the procedure of the systems of numeration which are taught at school mostly through exercise and usually remain unexpressed. Such pedantic rehearsal of decimal and binary numerals, however, is helpful to say something non-pedantic: all systems of numeration appear to be algorithmic by constitution. As any word implies a grammar, any number hides an algorithm – that is, a procedure for representing quantities and for performing operations with quantities. In conclusion, all numbers are *algorithmic numbers* as they are manufactured by those algorithms that are the systems of numerations. Numerals count nothing (so to speak); they are simply position holders in a procedure – an algorithm – of quantification.

The mechanisation of the algorithm

Algorithms for hand calculation were mechanised gradually. In seventeenth-century Europe, natural philosophers such as Pascal and Leibniz designed hand calculators to automate the four basic operations with the decimal system. These devices were not at all cabinet curiosities but signalled more profound epistemic changes. At the time, modern thought had already developed in close relation to machines, to the point that mechanical thinking can be recorded having an influence on philosophical thinking too. Descartes's famous 'Method' of reasoning, for instance, looked quite 'mechanical' in its emphasis upon the decomposition of a problem into simpler elements. According to the Polish economist Henryk Grossmann, it was not by accident that Descartes conceived his rational method while designing tooling machines himself. But Grossmann noted also a more profound relation between mathematics and machines: 'every mathematical rule has [a] mechanical character that spares intellectual work and much calculation.' This economic principle – to save time, work, and resources – remains a key aspect of algorithmic thinking and practices as they have been illustrated so far.[42]

As the following chapter will show, in the context of the industrial economy of the early nineteenth century, the first computing algorithm to be mechanised was Gaspard de Prony's method of difference to calculate large logarithmic tables, which Charles Babbage implemented in the Difference Engine. The Difference Engine was designed to embody only this type of algorithm, but Babbage's envisioned also a programmable machine – the Analytical Engine – which could express different kinds of equations (although it was never realised). The first computer algorithm or 'program' is considered to be Ada Lovelace's 'diagram for the computation of Bernoulli numbers' that was tentatively written for the Analytical Engine. Babbage's calculating engines represent the point of convergence of calculation algorithms and industrial automation, although they severely struggled among other difficulties, to represent the decimal system in mechanical gears.[43]

---

42 Henryk Grossmann, 'Descartes and the Social Origins of the Mechanistic Concept of the World', in *The Social and Economic Roots of the Scientific Revolution: Texts by Boris Hessen and Henryk Grossmann*, Berlin: Springer, 2009, 181.

43 See chapter 2.

In the twentieth century, algorithms for calculation were successfully automated thanks to the flexibility of the binary system.[44] Binary numerals are much easier to implement in an electric device than decimals into a mechanism, because an electric current's status that is on or off can directly represent the digits 0 and 1. In this way, the execution of addition and subtraction, for instance, is extremely simplified. Technically, binary operations started to be adopted and encoded in electric machines following the 1938 publication of US mathematician Claude Shannon's master's thesis 'A Symbolic Analysis of Relay and Switching Circuits'.[45] Shannon proposed for the first time to use the binary properties of electrical switches to represent not simply binary numbers and their operations, but propositional logic and, specifically, the Boolean logical operators AND, OR, and NOT.

After World War II, the binary code, the von Neumann architecture, and the engineering of efficient logic gates in microchips made possible the construction of fast computers and the formalisation of computer algorithms of larger size and higher complexity. For the first time in history, sequences of numerals came to represent not just quantities but instructions.[46] The so-called 'computer revolution' was not just about the use of *binary numerals* (binary digits, or bits) to encode human language and analogue content (digitisation) but about accelerating mechanical computation through *binary logic* (or Boolean logic). Contrary to the common view that stresses only the separation of hardware and software, digital computing is actually the imbrication, in the same medium of information *and* instruction, of binary numerals *and* Boolean logic – one as a complementary form of the other. In other words, with digital computing, the algorithm of numeration (binary numerals) and the algorithm of calculation (binary logic) have almost become one and the same thing.

In the digital age, the algorithm has risen to the role of an *abstract machine* (under the different denominations of program, software, code,

---

44 Within the Western tradition, Leibniz, inspired by the Chinese I Ching, already suggested binary numeration in his 1689 text *Explication de l'Arithmétique Binaire*.

45 Claude Shannon, 'A Symbolic Analysis of Relay and Switching Circuits', *Transactions of the American Institute of Electrical Engineers* 57, no. 12 (1938): 713–23.

46 Gödel introduced the idea of using numbers to represent mathematical functions (*Gödel numbering*) in his famous 1931 incompleteness theorems. Kurt Gödel, 'Über formal unentscheidbare Sätze der Principia Mathematica und verwandter Systeme I', *Monatshefte für Mathematik und Physik* 38 (1931).

The Material Tools of Algorithmic Thinking

and so forth), which is used to control electronic computing machines. As mentioned at the beginning of this chapter, the definition of 'algorithm' which is the most familiar in contemporary times is the one of computer science: 'a finite procedure of step-by-step instructions to turn an input into an output, independently of the data, and making the best use of the given resources'.[47] The abstraction of logic from content is one of the key aspects of technical and cognitive development: as with other techniques of abstraction, an algorithm has to operate independently of environmental constraints and the origin of data. This chapter has questioned, however, this reading of abstraction as separation from the world and its historical developments. In fact, the advent of machine learning has turned this static definition of algorithm upside down: machine learning algorithms have became adaptive, and from rigid sets of rules now they 'learn' rules from data.

The canonical definition describes the algorithm as the application of rigid rules, top down, on some input data. Data do not affect the behaviour of the algorithm: they are simply *passive information* to be processed by rules. On the contrary, machine learning algorithms change their internal rules (called parameters) according to the input data. As such, data are no longer passive, so to speak, but become *active information* that influences the parameters of the step-by-step procedure which is, then, no longer strictly predetermined by the algorithm. The breakthrough of machine learning is exactly about this shift: algorithms for data analytics become dynamic and change their rigid inferential structure to adapt to further properties of data – usually logical and spatial relations. The canonical example is an artificial neural network for pattern recognition that changes the parameters of its nodes according to the relations among the elements of the visual matrix. In this respect, the structure of the most recent AI algorithms is not different and distant from ancient mathematical practices that emerged by the continuous imitation of configurations of space, time, labour, and social relations.

---

47 Chabert mentions another definition of algorithm by Robert McNaughton that can be used as example of the technical ossification of the social processes previously illustrated: 1). The algorithm must be capable of being written in a defined language. 2). The question that is posed is determined by some input data, called enter. 3). The algorithm is a procedure which is carried out step by step. 4). The action at each step is strictly determined by the algorithm. 5). The output or answer (called exit) is clearly specified. Chabert, *A History of Algorithms*, 455.

As the historian of science Jürgen Renn has noted, after Damerow, machine learning algorithms are nothing 'superhuman' but part of the cycle of internalisation and externalisation of cognitive functions that belongs to all cultural techniques:

> After all, machine learning algorithms . . . are simply a new form of the externalization of human thinking, even if they are a particularly intelligent form. As did other external representations before them, such as calculating machines, for example, they partly take over – in a different modality – functions of the human brain. Will they eventually supersede and even displace human thinking? The crucial point in answering this question is not that their overall intelligence still lags far behind human and even animal intelligence, but that they can play out their full potential only within the cycle of internalization and externalization that . . . is the hallmark and driving force of cultural evolution.[48]

In a similar way, this introductory chapter served to see the algorithm concept in perspective – in its historical context as well as in the long evolution of knowledge systems. In short, it was, firstly, the mercantile acceleration of the late Middle Age, and, secondly, the rise of the information society that contributed to formalise the algorithm as it is known today. For a linguistic coincidence, the medieval term *algorismus* marked the passage from the additive to the positional system of numeration, while the recent use of the term 'algorithm' has marked the passage from decimal to binary numerals. These were not simply formal and technical shifts but also economic ones; after all, Hindu–Arabic numerals and algorithms for hand calculation were adopted to simplify accounting and mercantile transactions, while binary numerals were adopted because they could be implemented in electrical circuits and logic gates to accelerate industrial automation and state administration. Just as the first transition is related to early mercantilism, so is the second to industrial capitalism – particularly in its demand to speed up communication technologies and automate mental labour.[49]

---

48 Renn, *The Evolution of Knowledge*, 398.
49 See also Matteo Pasquinelli, 'From Algorism to Algorithm: A Brief History of Calculation from the Middle Ages to the Present Day', *Electra* 15 (Winter 2021–22): 93–102.

PART I
The Industrial Age

2
Babbage and the Mechanisation of Mental Labour

> *We must remember that another and a higher science, itself still more boundless, is also advancing with a giant's stride ... It is the science of calculation – which becomes continually more necessary at each step of our progress, and which must ultimately govern the whole of the applications of science to the arts of life.*
>
> Charles Babbage, *On the Economy of Machinery and Manufactures*, 1832[1]

Computation as division of labour

In early nineteenth-century England, 'computer' was not the name of a machine but of a human – namely an *office clerk*, often a woman, who had to make tedious calculations by hand for the government, the Astronomical Society, or the Navy. At times 'computers' were also working from home, receiving stacks of numbers to calculate and sending back results by mail: this was literally the first historical occurrence of a *computing network* that took the form of domestic labour and probably involved further family members. With the aim of streamlining this time-consuming and error-prone process, the polymath Charles

---

1 Charles Babbage, *On the Economy of Machinery and Manufactures*, London: Charles Knight, 1832, 316.

Babbage had the idea of replacing the repetitive work of many 'computers' with an automated machine powered by steam. Henry Colebrooke, presenting Babbage with a gold medal at the Astronomical Society of London in 1823 for the invention of the Difference Engine, declared:

> In other cases, mechanical devices have substituted machines for simpler tools or for bodily labour . . . But the invention to which I am adverting . . . substitutes mechanical performance for an intellectual process . . . Mr. Babbage's invention puts an engine in place of the [human] computer.[2]

Babbage's Difference Engine, celebrated as the precursor of modern computers, was born out of a business ambition – to automate the calculations of logarithms and sell error-free logarithmic tables, which were crucial in astronomy and for maintaining British hegemony in maritime trade. Among other instigators, it was the problem of the longitudinal calculus in open sea that gave a special impetus to mechanised computation. Small mechanical calculators already existed, but they were not automated and solved only the basic mathematical operations. Babbage had the idea of connecting a complex logarithmic calculator to the continuous motion provided by steam engines, so as not to have just a calculating device, but a *calculating engine* that could establish the business of calculation at an industrial scale – with the fantasies of unbounded performance and unfettered economic growth that the novel word 'engine' carried at the time. The idea of the automatic computer, in the contemporary sense, emerged out of the project to mechanise the mental labour of clerks rather than the old alchemic dream of building thinking automata – although the latter narrative would often be used, in the nineteenth century as much as in the century of corporate AI, to masquerade the former business.[3]

Precisely what kind of 'intellectual process', or mental labour, was

---

[2] Quoted in Simon Schaffer, 'Babbage's Intelligence: Calculating Engines and the Factory System', *Critical Inquiry* 21, no. 1 (1994): 203.

[3] For an animistic genealogy of thinking automata in the modern age, see Minsoo Kang, *Sublime Dreams of Living Machines: The Automaton in the European Imagination*, Camrbidge, MA: Harvard University Press, 2011. For a theological genealogy of machine design, see Ansgar Stöcklein, *Leitbilder der Technik; Biblische Tradition und technischer Fortschritt*, Munich: Moos, 1969.

Babbage aiming to mechanise? If we are to understand the limitations and potentialities of computation, this is a key clarification, without which even the definition of AI itself can only amplify misunderstandings. The first kind of mental labour to be mechanised was *hand calculation* – a specific skill that persisted until the model of the Turing machine, which was envisioned itself in the form of a human typist (a 'computer') reading and writing figures on a tape, as in a telegraph station. As chapter 9 will show by following a different genealogy of computation and AI, this was not to be the case with artificial neural networks for pattern recognition, which aimed to automate not hand calculation but the labour of perception and supervision.

Babbage's Difference Engine was a peculiar artefact. It was not a computer in the contemporary sense, because it did not distinguish software from hardware, instruction from information (fig 2.1). As it was at the same time both hardware *and* software, the Difference Engine appears aesthetically intriguing to contemporary eyes: its brass gears and rotating cylinders physically embodied a single algorithm, French mathematician Gaspard de Prony's 'method of differences', which was used to abbreviate the calculation of square numbers and logarithms. The Difference Engine was also not a computer in the contemporary sense, because it was not a programmable device: the title of an industrial machine featuring an independent input for information belongs to the more modest Jacquard loom.[4] The Difference Engine prototype was never finalised, while the Jacquard loom was produced in thousands of exemplars and became a driver of the industrial age. The Jacquard loom set a standard for information storage – the punched card – which IBM would maintain with little to no variation until the twentieth century.[5] Moreover, the first 'digital picture' – that is, an image described by a numerical file – happened

---

4 According to Frederick Pollock, mechanisation refers to the autonomisation of the energy source, while automation implies the independent role of information in the production process. The Jacquard loom was then an example of automation. See Frederick Pollock, *Automation: A Study of Its Economic and Social Consequences*, New York: Praeger, 1957. The autonomisation of the information component took place already in ancient musical automata. See Siegfried Zielinski and Eckhard Fürlus (eds), *Variantology 4: On Deep Time Relations of Arts, Sciences, and Technologies in the Arabic-Islamic World and Beyond*, Cologne: Walther König, 2008.

5 IBM punched cards were also used in the census of Jews in Nazi Germany. See Edwin Black, *IBM and the Holocaust: The Strategic Alliance between Nazi Germany and America's Most Powerful Corporation*, Washington: Dialog Press, 2001.

to be another textile artefact: an 1839 portrait of Jacquard himself that was woven using 24,000 of these punched cards.[6] Babbage kept a copy of Jacquard's portrait in his studio and adopted the punched card as an input format for another unrealised prototype – the Analytical Engine – whose design, unlike its precursor, theoretically separated information from instruction and could evaluate different types of equation.

The Difference Engine was not merely the invention of Babbage's lone speculative mind. As Simon Schaffer has noted, 'places of intelligence' across England assisted Babbage's experiments with mechanical computation and were ultimately the source of his 'machine intelligence'.[7] Schaffer remarks that Babbage had a more intimate relation with the industrial workshops as a locus of knowledge than with the universities, which, at the time, offered only conservative and notional curricula. Whereas the hagiographies still depict him as a solitary genius, Babbage was in fact deeply engaged in the industrial milieu of the age and in the debates of the emerging discipline of political economy. In fact, he authored one of the most influential industrial manuals of the time: *On the Economy of Machinery and Manufactures* (1832).

That the applied division of labour, rather than abstract mathematics, is the 'inventor' of automated computation is also confirmed by the opening of Babbage's book: 'The present volume may be considered as one of the consequences that have resulted from the Calculating-Engine, the construction of which I have been so long superintending.'[8] This is historical evidence that, as an expression of the division of labour, computation watched over the unfolding of industrial capitalism from its very outset, rather than being a product of its latest developments. While Babbage tried to convince his reader that the first manual ever published on the management of industrial production was inspired by the project of automated computation, a materialist historian would scrutinise such auto-mythography. Was it not, rather, the issue of labour organisation and insubordination which prompted the invention of new

---

6 Ada Lovelace, 'Notes' to Luigi Menabrea, 'Sketch of the Analytical Engine Invented by Charles Babbage', *Scientific Memoirs*, vol. 3, London: Richard and John E. Taylor, 1843.

7 Schaffer, 'Babbage's Intelligence', 204.

8 Babbage, *On the Economy of Machinery*, 1.

techniques of discipline and, therefore, urged Babbage to delve into the furnaces of industrial Europe?

Reckoning with clocks

The specific impetus for the mechanisation of mental labour and the invention of automated computation in England came from the need for precise logarithmic tables that, in an age of aggressive colonial expansion, were crucial to keeping orientation along maritime routes. Logarithmic tables, used to calculate the longitude in open sea, were highly unreliable because of human errors, which caused several shipwrecks and large commercial damages. The hagiographical anecdotes report that Babbage, mulling over the logarithm books and staring at their numerous errors, exclaimed: 'I wish to God these calculations had been executed by steam.'[9] The first project to accelerate the calculation of logarithmic tables, however, took place not in England but in France, where in 1791 the revolutionary government was engaged in reforming the official measuring system towards the metric system, investing in what Lorraine Daston (to draw an analogy with today's 'big data') calls 'big calculation'.[10] Pursuing an ambitious plan to make the decimal system the standard for angular measurements, the government asked Gaspard de Prony to design the division of the square angle with 100 rather than 90 degrees – a project which required the logarithmic translation of the old radial fractions into new ones. Though the plan of angular reform failed, and the millennia-old Sumerian partition of time remains a global standard to this day, the attempt would give momentum to the birth of automated computation.

The Scottish economist Adam Smith wrote a famous account of the division of labour in pin making in *The Wealth of Nations*. Smith's picture of the division of labour inspired de Prony, who designed accordingly a sort of *collective algorithm* for the calculus of

---

9 Babbage in November 1839, recalling events in 1821; quoted in Harry Wilmot Buxton, *Memoir of the Life and Labours of the Late Charles Babbage*, ed. Anthony Hyman, Cambridge, MA: MIT Press, 1987.

10 Lorraine Daston, 'Calculation and the Division of Labor, 1750–1950', *Bulletin of the German Historical Institute* 62 (Spring 2018).

logarithms. De Prony conceived a workflow that was organised as a social pyramid: at the top, he placed a class of mathematicians who would formulate the problem and pass it on to a second class of 'algebraists'; they would then prepare simple operations and data for a third class of human computers who would perform all of the actual calculations on paper sheets, then send them back to their superiors (see fig. 2.2). Students, often women, and sometimes 'a large number of unemployed hairdressers were used to fill out the numbers on the sheets by adding and subtracting'.[11] De Prony's algorithm applied the aforementioned method of differences, which is based on the fact that the difference between the squares of consecutive numbers remains constant and the interpolation of following squares can be easily reached by simple addition and subtraction in place of complex multiplication.[12]

Babbage had the idea of replacing the third class of workers of the calculating pyramid with a machine, as they were repeating tedious tasks of additions and subtractions of a simple difference. Eventually, the method of differences would provide the algorithm and the name for Babbage's machine: the Difference Engine. As mentioned above, at the time, mechanical calculators for basic operations already existed and were propelled by hand. Babbage had the idea of implementing this specific algorithm into a mechanical device and applying a steam engine as a source of motion to turn the calculation of logarithmic tables into an industrial business of scale. Once the Difference Engine was set in motion, it was supposed to calculate a whole logarithmic table without stopping. Babbage's project was a fascinating contrivance that sought to give unbounded computational power to cogs and wheels made of brass and wood. Today, the use of steam as a source of energy for calculation may endure only in the science fiction genre steampunk, but in Babbage's time it was a venture into a world very different than the one we currently inhabit – one where automated computation would run without electricity.

---

11 Ivor Grattan-Guinness, 'Charles Babbage as an Algorithmic Thinker', *IEEE Annals of the History of Computing* 3 (1992): 40.

12 In fact, logarithmic and trigonometrical functions do not always maintain a constant difference continuously. The method of differences, then, is a heuristic approximation that is valid only for specific ranges.

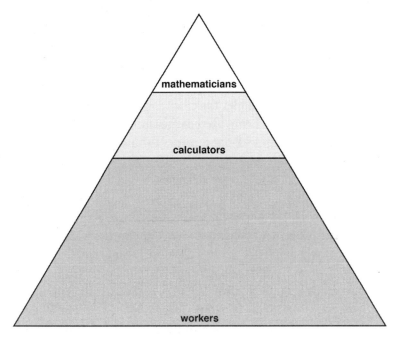

Figure 2.1. Scheme for the implementation of de Prony's algorithm as division of labour. Lorraine Daston, 'Calculation and the Division of Labor, 1750–1950', *Bulletin of the German Historical Institute* 62 (Spring 2018): 11.

Babbage's first prototype of the Difference Engine, modestly, was still propelled by hand. Interestingly, the first device that would take the role of translating de Prony's algorithm for hand calculation into 'matter' was a familiar one: the clock (see fig. 2.2). Babbage published the general concept of the Difference Engine in the often-forgotten chapter, 'On the Division of Mental Labour', in his 1832 book. There, he proposed 'what may, perhaps, appear paradoxical to some of our readers – that the division of labour can be applied with equal success to mental operations, and that it ensures, by its adoption, the same economy of time'.[13] Following de Prony's method of differences, Babbage deconstructed the calculation of logarithmic tables in modular steps and implemented them into a new mechanical algorithm. The three columns for the table of the method of differences were represented by clocks that Babbage subsequently implemented as rotating cylinders. In the first working

---

13 Babbage, *On the Economy of Machinery*, 153.

prototype of the Difference Engine (ca. 1833), the step-by-step rotation of cylindrical 'clocks' replaced the movements of a hand, adding digits on a piece of paper. The artefact of the clock had here the exemplary role of heuristic mediator between the system of numeration and the algorithm of calculation. If automation is pursued out of the need to save time, its implementation under the clock form itself is emblematic. It is also revealing that, after being used to measure manual labour productivity in the factory, the clock hand comes to automate hand calculation itself.

These cylindrical clocks were designed to receive a number as an incremental rotation, to add it to a previous number of increments, and to perform a total output under the form of a further incremental rotation. This movement was, however, imprisoned within a mechanism that could perform, irreversibly, only one large, continuous operation. Just as it did not distinguish hardware from software, Babbage's clock-cylinders also did not really distinguish numbers from processes, or memory from operations. These two functions were to be separated in the design of the Analytical Engine (and later in modern computers as the division between memory and the Central Processing Unit). Another limitation that Babbage's mechanical algorithm confronted was the decimal system itself and the problem of automating the carryover of the tens place, which afflicted mechanical calculation since the time of Pascal.[14] It must be remembered that the binary system was implemented (thanks to Leibniz, Boole, Turing, Shannon, and von Neumann, among others) because it technically simplifies addition and subtraction. An electric switch can turn on and off, with these two states representing all the necessary units of numeration (see chapter 6). The Difference Engine's wheels, on the other hand, struggled to contain ten numerals, and Babbage tried to resolve the problem of the remainder by a sophisticated yet clumsy carriage return.

As Matthew Jones has illustrated, for Babbage as for many philosopher-inventors of the modern age, the enterprise of mechanical calculation was not to be distinguished from that of natural

---

14 For the century-long problem of mechanising carry, see Matthew L. Jones, *Reckoning with Matter: Calculating Machines, Innovation, and Thinking about Thinking from Pascal to Babbage*, Chicago: University of Chicago Press, 2017.

philosophy, which aimed at 'reckoning with matter', for the precise reason that reckoning was considered a lower mental activity, something that the 'mechanical' classes (or their machine equivalent) had to perform for the upper classes. Indeed, most modern natural philosophers (Hobbes aside) maintained that the mind could not be reduced to mechanism.[15] In this political climate, mental labour could therefore be automated because it was a task of the working class, and not one to be regarded as 'thinking' proper.

Principles of labour analysis

Although the Difference Engine easily evokes fascination for historians of Victorian science and technology, Babbage should be remembered less for the machine itself than for the principles of the division of labour that inspired its design. Historians of science such as Daston and Schaffer have contributed to questioning the Difference Engine's status as the solo violin of early automated computation and instead made visible, on the stage of the industrial age, a less seductive yet more logical protagonist: the division of labour and its social hierarchy. Schaffer has highlighted that Babbage's 'machine intelligence' proceeded from the 'mindful hands' of workers, craftsmen, and machinists who were building experimental contrivances, as seen above, in 'places of intelligence' such as workshops and factories rather than royal academies. As Jones details, Babbage publicly pursued Francis Bacon's 'hope for the discovery of a philosophical theory of invention'.[16] Yet the secret of his calculating engine was not the imitation of God's foresight (as Babbage argued)[17] so much as the everyday business of these workshops and factories, made of continuous failures and conflicts with workers, including the insubordination of Babbage's own team. In order to better understand the design of Babbage's machines and their quality of 'machine intelligence', it is therefore necessary to explicate his two principles of labour analysis: (1) the *labour theory of the machine*, which states that a new machine comes to imitate and replace a previous

---

15 See Jones, *Reckoning with Matter*; Daston, 'Calculation and the Division of Labor'.
16 Jones, *Reckoning with Matter*, 1.
17 Charles Babbage, *Ninth Bridgewater Treatise*, London: John Murray, 1838.

Repetitions of Process.	Movements.	Clock A. Hand set to I.	Clock B. Hand set to III.	Clock C. Hand set to II.
		TABLE.	First difference.	Second difference.
1	Pull A.	A. strikes 1	. . . .	. . . .
	— B.	The hand is advanced (by B.) 3 divisions . .	B. strikes 3	. . . .
	— C.	. . . .	The hand is advanced (by C.) 2 divisions . .	C. strikes 2
2	Pull A.	A. strikes 4	. . . .	. . . .
	— B.	The hand is advanced (by B.) 5 divisions . .	B. strikes 5	. . . .
	— C.	. . . .	The hand is advanced (by C.) 2 divisions . .	C. strikes 2
3	Pull A.	A. strikes 9	. . . .	. . . .
	— B.	The hand is advanced (by B.) 7 divisions . .	B. strikes 7	. . . .
	— C.	. . . .	The hand is advanced (by C.) 2 divisions . .	C. strikes 2
4	Pull A.	A. strikes 16	. . . .	. . . .
	— B.	The hand is advanced (by B.) 9 divisions . .	B. strikes 9	. . . .
	— C.	. . . .	The hand is advanced (by C.) 2 divisions . .	C. strikes 2
5	Pull A.	A. strikes 25	. . . .	. . . .
	— B.	The hand is advanced (by B.) 11 divisions .	B. strikes 11	. . . .
	— C.	. . . .	The hand is advanced (by C.) 2 divisions . .	C. strikes 2
6	Pull A.	A. strikes 36	. . . .	. . . .
	— — B.	The hand is advanced (by B.) 13 divisions .	B. strikes 13	. . . .
	— C.	. . . .	The hand is advanced (by C.) 2 divisions . .	C. strikes 2

Figure 2.2. Design for the implementation of de Prony's algorithm into a mechanism. Charles Babbage, *On the Economy of Machinery and Manufactures*, London: Charles Knight, 1832, 161.

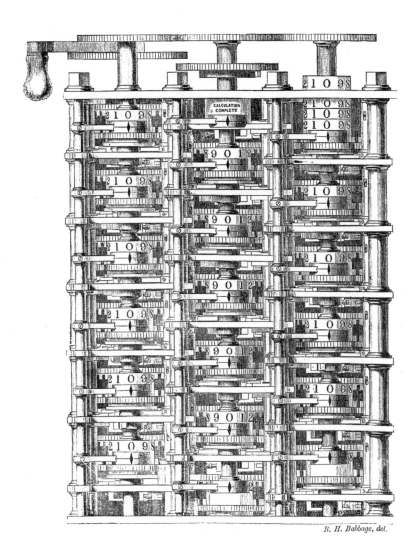

Figure 2.3. Babbage's Difference Engine. Charles Babbage, *Passages from the Life of a Philosopher*, London: Longman, Roberts & Green, 1864, front cover.

division of labour; and (2) the *principle of labour calculation* (usually called the 'Babbage principle'), which states that the division of labour allows for the calculation and purchase of exactly the quantity of necessary labour.

As the founder of modern economics, Adam Smith was the first to have sketched a labour theory of the machine in *The Wealth of Nations* (1776) by recognising that new machines are 'invented' by imitating the organisation of tasks in the workplace: 'The invention of all those machines by which labour is so much facilitated and abridged seems to have been originally owing to the division of labour.'[18] Whereas the independent tool emerges out of the repetition of a simple manual activity, the machine emerges out of assemblages of these tools. Given his greater technical experience, Babbage formulated this idea better in his *On the Economy of Machinery and Manufactures*:

> Perhaps the most important principle on which the economy of a manufacturer depends, is the *division of labour* amongst the persons who perform the work . . . The division of labour suggests the contrivance of tools and machinery to execute its processes . . . When each process has been reduced to the use of some simple tool, the union of all these tools, actuated by one moving power, constitutes a machine.[19]

The labour theory of the machine is based on a postulate at once technical and economic, according to which a machine emerges only after a coordination of tools has been tested and proved to be successful for production and cost reduction. If Smith and Babbage are right, and a machine emerges as the experimentation and implementation of a collective division of labour, a political issue comes straightaway to the fore: Who is actually the inventor of the machine? Who can claim credit for its

---

18 Adam Smith, *The Wealth of Nations* (1776), book 1, chapter 1. The passage continues: 'A great part of the machines made use of in those manufactures in which labour is most subdivided, were originally the inventions of common workmen, who, being each of them employed in some very simple operation, naturally turned their thoughts towards finding out easier and readier methods of performing it.' Smith adds that the invention of a new machine is due to its very users and to 'philosophers' (scientists and engineers, in the parlance of the time). See Tony Aspromourgos, 'The Machine in Adam Smith's Economic and Wider Thought', *Journal of the History of Economic Thought* 34, no. 4 (2012): 475–90.

19 Babbage, *On the Economy of Machinery*, 131–6.

invention? Workers, factory masters, engineers, or the orchestration of all these actors? Who owns the right over such a collective division of labour? These were actual and highly debated issues in the so-called 'Machinery Question' of the nineteenth century (see chapter 3).

Babbage's further contribution was to frame the *labour theory of the machine* – namely, that a machine imitates and replaces a previous division of labour – in terms of economic planning. In fact, the division of labour itself emerged not just to better organise labour in modular tasks but to precisely measure (to compute, one is tempted to say) the *cost* of each task. The so-called Babbage principle is canonically formulated in this passage:

> The master manufacturer, by dividing the work to be executed into different processes, each requiring different degrees of skill and force, can purchase exactly that precise quantity of both which is necessary for each process; whereas, if the whole work were executed by one workman, that person must possess sufficient skill to perform the most difficult, and sufficient strength to execute the most laborious, of the operations into which the art is divided.[20]

The Babbage principle states that the organisation of a production process into small tasks (the division of labour) allows for the calculation and precise purchase of the quantity of labour that is necessary for each task (the division of value). The division of labour establishes a privileged perspective for the surveillance of labour, but also helps to modulate the extraction of surplus labour from each worker according to need. In more analytical terms, the Babbage principle posits that the abstract diagram of the division of labour helps to organise production while at the same time offering an *instrument* for measuring the value of labour. In this respect, the division of labour provides not only the design of machinery but also of the business plan.

A fundamental theory of automated computation came from Babbage's application of his principles of labour calculation to the division of mental labour. Notably, he already saw the factory as a sort of knowledge economy from which to extract exact 'quantity of skill and knowledge which is required' from each worker:

---

20 Ibid., 137.

The effect of the division of labour, both in mechanical and in mental processes, is, that it enables us to purchase and apply to each process precisely that quantity of skill and knowledge which is required for it: we avoid employing any part of the time of a man who can get eight or ten shillings a day by his skill in tempering needles, in turning a wheel, which can be done for sixpence a day; and we equally avoid the loss arising from the employment of an accomplished mathematician in performing the lowest processes of arithmetic.[21]

Combining both Babbage's principles, one could say that computation emerged as both the *automation* of the division of mental labour and the *calculus of the costs* of such labour. One could postulate that, under the logic of computation, the automation of labour and the calculus of the cost of labour even become the same thing. After all, to compute means to measure the costs of labour in terms of time, space, energy, resources, and capital. This often neglects, from capital's perspective, the 'human cost' of such labour. As the historian of science Norton Wise has remarked, the division of labour

is actually one of the hierarchy of labour, rather than merely division . . . It separates skill from brute force in order that the manufacturer does not have to pay for them simultaneously . . . But the single principle applies to the entire hierarchy, to machines as to human labourers and to mental as to physical labour, requiring that the numbers and kinds of all sources be allocated so as to minimize cost of production. It is a principle of the interior organization of a factory . . . which Babbage sought to generalize to the entire political economy.[22]

What Babbage's principles of labour analysis implied was the further discrimination between skilled and unskilled workers and the 'automation', ultimately, of social hierarchies of knowledge. In conclusion, Babbage's labour theory of the machine is of

---

21 Ibid., 162.
22 Norton Wise and Crosbie Smith, 'Work and Waste: Political Economy and Natural Philosophy in Nineteenth Century Britain' (part 2), *History of Science* 27, no. 4 (1989): 411. 'Hierarchy' as collective noun is in the original.

extraordinary importance when it is combined with his principle of labour calculation: together, they seem to define the industrial machine not just as a productive apparatus but also as an instrument of measurement of labour. Ultimately, the Babbage principle represents a *machine theory of value* – that is, a model to mechanically represent and compute labour costs and capital investments. In a highly formalised way, it can be said that the labour theory of the machine and the machine theory of value together form a techno-economic principle according to which the machine is built by the division of labour in order to achieve a more accurate calculation and extraction of surplus value.

Analytical intelligence and machine semiotics

Mulling over broken tools, uneven cogs, and unfinished machines, Babbage found himself in a situation not uncommon to many other inventors: he needed to codify an artificial language in order to improve and accelerate design. In *On the Economy of Machinery*, he writes:

> It is possible to construct the whole machine upon paper, and to judge of the proper strength to be given to each part as well as to the framework which supports it, and also of its ultimate effect, long before a single part of it has been executed. In fact, all the contrivance, and all the improvements, ought first to be represented in the drawings.[23]

As Jones has noted, Babbage's analytical hopes soon clashed with the contingencies of implementation and the necessity of human cooperation. In his autobiography, Babbage himself admitted that 'draftsmen of the highest order were necessary to economize the labour of my own head; whilst skilled workmen were required to execute the experimental machinery to which I was obliged constantly to have recourse'.[24] Babbage's project of a 'machine semiotics' (as Schaffer has called it) extended his principles of labour analysis and

---

23 Babbage, *On the Economy of Machinery*, 207.
24 Charles Babbage, *The Works of Charles Babbage*, vol. 11, ed. Martin Campbell-Kelly, London: Pickering, 1989, 85.

expressed an intuition similar to what more recent authors have alternatively defined as 'mechanical thinking', 'computational thinking', or 'algorithmic thinking'.[25]

After basing the calculating engines on the analysis of the division of mental labour, Babbage tried to establish a notational system for machine design on similar principles. In order to better articulate their logical form, the design of the calculating engines called for a symbolic metalanguage (a second-order representation), which Babbage termed 'mechanical notation'. Babbage expressed this project in two texts: 'On a Method of Expressing by Signs the Action of Machinery' (1826) and 'Laws of Mechanical Notation' (1851).[26] The epistemic dimension of machine making was already clear to Babbage in another note from 1851:

> It is not a bad definition of man to describe him as a tool-making animal. His earliest contrivances to support uncivilized life were tools of the simplest and rudest construction. His latest achievements in the substitution of machinery, not merely for the skill of the human hand, but for the relief of the human intellect, are founded on the use of tools of a still higher order.[27]

The purpose of mechanical notation was to represent dynamic diagrams of machine states, over and beyond the traditional static drawings. Considering the nature of these calculating machines, Babbage's mechanical notation can be considered the embryonic stage of what would later become the flow charts and programming languages of twentieth-century digital computers. The logical equivalence between

---

25 For 'machine semiotics', see Simon Schaffer, 'Babbage's Intelligence: Calculating Engines and the Factory System', *Critical Inquiry* 21, no. 1 (1994): 207. On mechanical thinking, see the work of Department 1 at the Max Planck Institute for the History of Science, Berlin, under the direction of Jürgen Renn. For instance: Rivka Feldhay et al. (eds), *Emergence and Expansion of Preclassical Mechanics*, vol. 270, Berlin: Springer, 2018. On the now-popular expression 'computational thinking', see Jeannette Wing, 'Computational Thinking', *CACM Viewpoint* 49, no. 3 (March 2006).

26 Charles Babbage, 'On a Method of Expressing by Signs the Action of Machinery', F.R.S. Philosophical Transactions, 16 March 1826; 'Laws of Mechanical Notation', July 1851. See also Henry P. Babbage 'Mechanical Notation, exemplified on the Swedish Calculating Machine', Glasgow: British Association, September 1855.

27 Charles Babbage, *The Exposition of 1851*, London: John Murray, 1851.

Babbage's mechanical notation and digital computer language is no coincidence: in fact, it is possible for the latter to emulate the former.[28]

The idea of mechanical notation must also be contextualised as part of the intellectual milieu of Babbage's time – and in particular, the controversy about the rise of mathematical analysis in British universities against the traditional curricula of geometry. In the early nineteenth century, a dispute broke out at Cambridge University between geometry scholars proud of their practical insights and the new *algebraists*. The latter were accused of merely adopting a fashionable pose from France – called 'Analysis' – which mired them in abstractions lacking any practical use. With industrial capitalism storming through the United Kingdom, however, the arrival of 'analytical' thinking in academia can be considered as echoing another equally urgent form of 'analysis' for the time: the *analysis of labour*. Indeed, mathematical analysis appeared to be encouraged by the economic demand for more analytical intelligence on labour and machines. As will be shown in chapter 4, even Marx adopted Babbage's terminology, writing in the *Grundrisse* that the best method for designing machines is not the application of science (the 'analysis of nature') to industry but the 'analysis through the division of labour'.[29]

Regarding the process of invention, Babbage suggested a threefold design: (1) the analysis of the machine's design through its components; (2) the simplification of the machine's design; and (3) the implementation of the simplified design, which usually brings further adjustments.[30] The algorithmic nature of this method is manifest in the centrality given to the simplification process, which makes it equivalent to a method of optimisation and economisation of resources. In his late autobiography, *Passages from the Life of a Philosopher*, Babbage remarked, unsurprisingly, that the economy of time – which is key to the division of labour and the design of machines – is also key to the calculation programs to be run on the Analytical Engine (a principle known today as *algorithmic*

---

28 Doron Swade has formalised Babbage's notation into a proper computer language. See Doron Swade, 'Project Report to Computer Conservation Society Committee', London, 19 March 2015.

29 Karl Marx, *Grundrisse: Foundations of the Critique of Political Economy*, London: Penguin 1973, 704.

30 Mark Priestley, *A Science of Operations: Machines, Logic, and the Invention of Programming*, Berlin: Springer, 2011, 29.

efficiency): 'As soon as an Analytical Engine exists, it will necessarily guide the future course of the science. Whenever any result is sought by its aid, the question will then arise – By what course of calculation can these results be arrived at by the machine in the *shortest time*?'[31] Babbage was therefore not just the mathematician that most of his biographers portray, but already an 'algorithmic thinker', as the principle of design optimisation and resource economisation was key to his (uncompleted) machines as much as it was to his (speculative) algorithms.[32]

Babbage's mechanical notation grew, essentially, out of the analysis of labour. If the design of the labour process forges the machine's design, then, in a similar way, the machine's design itself inspires the machine language as a second-order representation. As elucidated in chapter 1, the scaffolding of further levels of *abstraction* in industrial machine design is typical of the development of cultural techniques. Babbage's principles are specifically principles of the 'analytical intelligence' of labour that can also be useful for illuminating the history of AI in the following century, as a continuous implementation and automation of labour tasks.[33] In short: the analytical intelligence of labour is what grounds the analytical intelligence of the machine.

Ada Lovelace: When computers were women

Babbage was no solitary genius. Since the publication of Bertram Bowden's anthology *Faster than Thought* (1953), the figure of Ada Lovelace and her contribution to Babbage's projects have progressively been acknowledged.[34] Indeed, a growing literature today, while overlooking the role of other 'anonymised' women in the business of calculation of the time, celebrates her as the 'first woman programmer of history'.[35] Lovelace, the daughter of the poet Lord Byron and the often

---

31 Charles Babbage, 'Of the Analytical Engine', in *Passages from the Life of a Philosopher*, London: Longman, 1864, 137.

32 Ivor Grattan-Guinness, 'Charles Babbage as an Algorithmic Thinker', *IEEE Annals of the History of Computing* 3 (1992).

33 For the definition of 'analytical intelligence', see Daston, 'Calculation and the Division of Labor'.

34 Bertram Bowden, *Faster than Thought: A Symposium on Digital Computing Machines*, London: Pitman, 1953.

35 See Joan Baum, *The Calculating Passion of Ada Byron*, Hamden, CT, Archon

forgotten mathematician Anne Isabella Milbanke, was passionate about the algebraic notation known at the time as 'Analysis' – so much so that she even gave herself the futuristic title of 'Analyst'.[36] This passion for mathematics and abstract notation brought her to become an acquaintance of Babbage.

Lovelace assisted Babbage in designing the Analytical Engine and wrote the first-ever documented machine program. Even though the Analytical Engine was never realised, Lovelace's virtual program, understood as a set of instructions to be executed by a machine, is considered the first example of present-day algorithms – though she never used the term 'algorithm' herself but rather called the program a 'diagram'. Her 'Diagram for the computation of Bernoulli numbers' is found in the 'Notes' to Luigi Menabrea's 'Sketch of the Analytical Engine Invented by Charles Babbage'.[37] Menabrea (who would become Italian prime minister in 1867) met Babbage in Turin as a young mathematician and wrote an account of the Analytical Engine. Babbage subsequently asked Lovelace to translate Menabrea's text from French, and she expanded it with an appendix that turned out to be longer than the main text.

These 'Notes' are a milestone in the history of computation, as they sketch embryonic postulates of what in the twentieth century would become known as 'computer science' and what in her time Lovelace defined as the 'science of operations'.[38] Her intention was to distinguish between the logical and mechanical structure of the Engine – or between software and hardware, as one would say today. Taking the mechanical contrivances that made such logical powers possible as a substrate, she detailed the improvements upon its predecessor, the Difference Engine

---

Books, 1986; James Essinger, *A Female Genius: How Ada Lovelace Started the Computer Age*, London: Gibson Square Books, 2013; Doris L. Moore, *Ada, Countess of Lovelace: Byron's Legitimate Daughter*, New York: Harper & Row, 1977; Betty Alexandra Toole (ed.), *Ada, the Enchantress of Numbers: A Selection from the Letters of Lord Byron's Daughter and Her Description of the First Computer*, Mill Valley, CA: Strawberry Press, 1992.

36 While today we distinguish *symbolic analysis* (i.e., algebra) and *numerical analysis* (i.e., the study of algorithms) as they respond to different procedures of symbolisation and reasoning, in Lovelace's time the term 'analysis' connoted both algebraic notation and the differential calculus of Newton and Leibniz, as opposed to the established curricula of geometry.

37 Lovelace, 'Notes'.

38 Ibid., 22.

(which, as mentioned above, implemented just a single algorithm). It can therefore be said that Lovelace dedicated herself to the ambitious and complex task of describing the Analytical Engine as the first *general purpose computer*, as it would be termed today.

By envisioning a logical machine that could express all possible equations and their evaluation, Lovelace advanced a definition of 'operation' which was more general and universal than the operation upon numbers as it is understood by traditional mathematics. Her science of operations included the abstract manipulation of *any* entity, not just numbers, suggesting in this way a broader meaning also for the definition of automation. She wrote:

> It may be desirable to explain, that by the word *operation*, we mean *any process which alters the mutual relation of two or more things*, be this relation of what kind it may. This is the most general definition, and would include all subjects in the universe. In abstract mathematics, of course operations alter those particular relations which are involved in the considerations of number and space, and the *results* of operations are those peculiar results which correspond to the nature of the subjects of operation. But the science of operations, as derived from mathematics more especially, is a science of itself, and has its own abstract truth and value.[39]

In other words, Lovelace defined as an 'operation' the control of material and symbolic entities beyond the second-order language of mathematics (like the idea, discussed in chapter 1, of an algorithmic thinking beyond the boundary of computer science). In a visionary way, Lovelace seemed to suggest that mathematics is not the universal theory par excellence but a particular case of the science of operations. Following this insight, she envisioned the capacity of numerical computers qua *universal machines* to represent and manipulate numerical relations in the most diverse disciplines and generate, among other things, complex musical artefacts:

> [The Analytical Engine] might act upon other things besides number, were objects found whose mutual fundamental relations could be

---

39 Ibid., 22.

Babbage and the Mechanisation of Mental Labour

expressed by those of the abstract science of operations, and which should be also susceptible of adaptations to the action of the operating notation and mechanism of the engine... Supposing, for instance, that the fundamental relations of pitched sounds in the science of harmony and of musical composition were susceptible of such expression and adaptations, the engine might compose elaborate and scientific pieces of music of any degree of complexity or extent.[40]

Doron Swade, a historian of computing, has given the following compelling portrait of Lovelace, who, as a pioneer of 'general purpose computation', was discovering the potentiality of symbolic manipulation beyond the field of mathematics:

Ada saw something that Babbage in some sense failed to see. In Babbage's world his engines were bound by number... What Lovelace saw... was that number could represent entities other than quantity. So once you had a machine for manipulating numbers, if those numbers represented other things, letters, musical notes, then the machine could manipulate symbols of which number was one instance, according to rules. It is this fundamental transition from a machine which is a number cruncher to a machine for manipulating symbols according to rules that is the fundamental transition from calculation to computation – to general-purpose computation.[41]

Lovelace sensed the speculative horizons upon which the Analytical Engine, with its unbound powers of computation, would open:

The Analytical Engine does not occupy common ground with mere 'calculating machines'. It holds a position wholly its own; and the considerations it suggests are most interesting in their nature. In enabling mechanism to combine together general symbols in successions of unlimited variety and extent, a uniting link is established between the operations of matter and the abstract mental processes of

---

40 Ibid., 23.
41 Quoted in John Fuegi and Jo Francis, 'Lovelace and Babbage and the Creation of the 1843 "Notes"', *Annals of the History of Computing* 25, no. 4 (October–December 2003): 16–26.

the *most abstract* branch of mathematical science. A new, a vast, and a powerful language is developed for the future use of analysis, in which to wield its truths so that these may become of more speedy and accurate practical application for the purposes of mankind than the means hitherto in our possession have rendered possible. Thus not only the mental and the material, but the theoretical and the practical in the mathematical world, are brought into more intimate and effective connexion with each other. We are not aware of its being on record that anything partaking in the nature of what is so well designated the Analytical Engine has been hitherto proposed, or even thought of, as a practical possibility, any more than the idea of a thinking or of a reasoning machine.[42]

Lovelace's 'Notes' contain, however, the first dismissal of AI in a world that was already cultivating the anthropomorphic projection of a machine that could 'think' like a human. In the famous note 'G', she wrote:

The Analytical Engine has no pretensions whatever to *originate* anything. It can do whatever we *know how to order it* to perform. It can *follow* analysis; but it has no power of *anticipating* any analytical relations or truths. Its province is to assist us in making *available* what we are already acquainted with. This it is calculated to effect primarily and chiefly, of course, through its executive faculties; but it is likely to exert an *indirect* and reciprocal influence on science itself in another manner. For, in so distributing and combining the truths and the formulae of analysis, that they may become most easily and rapidly amenable to the mechanical combinations of the engine, the relations and the nature of many subjects in that science are necessarily thrown into new lights, and more profoundly investigated. This is a decidedly indirect, and a somewhat *speculative,* consequence of such an invention. It is however pretty evident, on general principles, that in devising for mathematical truths a new form in which to record and throw themselves out for actual use, views are likely to be induced, which should again react on the more theoretical phase of the subject. There

---

42 Charles Babbage, *Charles Babbage and His Calculating Engines,* New York: Dover Publications, 1961, 25.

are in all extensions of human power, or additions to human knowledge, various *collateral* influences, besides the main and primary object attained.[43]

That the Analytical Engine could follow analysis means that it could represent and embody the analytical construction of a problem as an algebraist could do. Moreover, that the Analytical Engine had 'no power of anticipating any analytical relations or truths' means that it could not exceed or break the chain of reasoning that it was representing and materially embodying – just as today's algorithms for data analytics that are rebranded as 'machine learning' and 'artificial intelligence' cannot creatively break the rules on which they are based and, more importantly, cannot consistently invent new ones.

Babbage reluctantly recognised Lovelace's contribution, asking to publish her notes on the Analytical Engine anonymously. For her resistance against Babbage's chauvinism, Lovelace is, without a doubt, an exemplary figure of technical curiosity and emancipation in an academic and scientific world dominated by men.[44] Yet her hagiographic portrait has also to be placed into its context. Both Babbage and Lovelace's stories belong to a narrative of the industrial era in which social hierarchies and intellectual debts are mystified by a predictable bourgeois personality cult. One example is a quote from Lovelace that has become a routine slogan for the digirati: 'We may say most aptly that the Analytical Engine weaves algebraic patterns just as the Jacquard loom weaves flowers and leaves.' To the romanticism of this quote, Schaffer contrasted a harsh observation: 'Lovelace never raised the problem of the substitution of weavers' intelligence by a series of automatic program cards nor the consequent sufferings of London's skilled unemployed.'[45]

---

43 Lovelace, 'Notes', 44.

44 See Jennifer Light, 'When Computers Were Women', *Technology and Culture* 40, no. 3 (1999): 455–83.

45 Simon Schaffer, 'Babbage's Dancer and the Impresarios of Mechanism', in *Cultural Babbage: Technology, Time and Invention*, ed. Francis Spufford and Jenny Uglow, London: Faber & Faber, 1996, 77.

The march of the material intellect

According to Babbage's vision, the calculating engines were also tools for the measurement, disciplining, and surveillance of labour – that is, for the 'intelligence' of labour, at least in the sense the word still carried at the time. According to Schaffer, in fact, 'in early nineteenth-century Britain the word intelligence simultaneously embodied the growing system of social surveillance and the emerging mechanisation of natural philosophies of mind.'[46] Well before the technocratic ambitions of cybernetics in the twentieth century, Babbage had cultivated a larger technocratic vision of society through his calculating machines. The hyperbolic publicist Dionysius Lardner professed that Babbage's new system of mechanical notation could be useful for describing and operating 'an extensive factory, or any great public institution, in which a vast number of individuals are employed, and their duties regulated'.[47]

In a similar vein of industrialist propaganda, Babbage's book on machinery and manufactures concluded with a chapter on the grandiose progress of British capitalism under the banner of 'abstract Science.'[48] There, Babbage specifically asserted that the accumulation of science does not follow the laws of scarcity of physical forces and material production but is virtuously amplified over time:

> Science and knowledge are subject, in their extension and increase, to laws quite opposite to those which regulate the material world. Unlike the forces of molecular attraction, which cease at sensible distances; or that of gravity, which decreases rapidly with the increasing distance from the point of its origin; the further we advance from the origin of our knowledge, the larger it becomes, and the greater power it bestows upon its cultivators, to add new fields to its dominions.[49]

Commenting on Babbage's ideology, and its striking resemblance to twentieth-century proclamations about the knowledge society, Norton

---

46 Schaffer, 'Babbage's Intelligence', 204.
47 Quoted in Priestley, *A Science of Operations*, 30. Dionysius Lardner, 'Babbage's Calculating Engine', *Edinburgh Review*, 59, 263–327 (1834). Reprinted in Charles Babbage, *The Works of Charles Babbage*, vol. 2, London: William Pickering, 1989.
48 Babbage, *On the Economy of Machinery*, 307.
49 Ibid., 315.

Wise writes:

> The engine metaphor now extends from the literal steam engine setting machinery in motion, to capital as the engine of labour, to the machine economy as a social engine, to scientific knowledge as an engine of practical action. Inevitably, scientific knowledge represented capital in the economy of knowledge, a reservoir of moving force which continued to accumulate at compound interest.[50]

These optimistic views about knowledge development were not uncommon at the time: as shown in the following chapter, Ricardian socialists such as William Thompson and Thomas Hodgskin had similar utopian theories about the accumulation of knowledge labour from the perspective of the workers' movement. According to Babbage, the ratio of the 'continually increasing field of human knowledge' is exponential, and one wonders (as did the Ricardian socialists and Marx) what the effect would be of such an overproduction of knowledge and science on the economy and on capital accumulation. It is at this climax of accumulation of knowledge and science that Babbage prophetically announced the hegemonic rise of a new science – the science of *calculation*:

> We must remember that another and a higher science, itself still more boundless, is also advancing with a giant's stride, and having grasped the mightier masses of the universe, and reduced their wanderings to laws, has given to us in its own condensed language, expressions, which are to the past as history, to the future as prophecy. It is the same science which is now preparing its fetters for the minutest atoms that nature has created: already it has nearly chained the ethereal fluid, and bound in one harmonious system all the intricate and splendid phenomena of light. It is the science of *calculation* – which becomes continually more necessary at each step of our progress, and which must ultimately govern the whole of the applications of science to the arts of life.[51]

---

50 Wise and Smith, 'Work and Waste', 414. See also Charles Babbage, 'Preface', in *Memoirs of the Analytical Society*, by Charles Babbage and John Herschel, Cambridge: Cambridge University Press, 1813, xxi.

51 Babbage, *On the Economy of Machinery*, 316.

Ideology here wins our over Babbage's scientific achievements. While in an earlier chapter of his book, Babbage grounded the science of calculation on the principles of labour analysis, he ultimately presented it as the materialisation of science in the abstract.

In a similar way to the contemporary AI discourse, Babbage captured the collective intelligence of the division of labour and instrumentalised it to build a technocratic view of society.[52] Babbage's rhetoric never explicitly acknowledged the Machinery Question – the public debate about workers who were being replaced by machines. Rather, for him, knowledge of steam power and the new science of automated calculation were meant to serve solely as a multiplier of productivity.[53] In the words of the historian William Ashworth: 'Babbage's work on his calculating machine was the march of the material intellect set to the rhythm of the factory.'[54]

---

52 On Babbage's political plan, Wise has commented: 'Babbage made knowledge, or mental labour, the source of transformation in the economy, and . . . he sought to bring its transformational operations within the purview of mechanical laws, just as he did in the analytical engine.' Wise and Smith, 'Work and Waste', 416.

53 Babbage's last thoughts turn not to the depletion of 'human resources' but of natural ones, repeating an argument that would become a leitmotiv in the following century: 'The source of this power is not without limit, and the coal-mines of the world may ultimately be exhausted.' Already in his own time Babbage recommended the conversion to geothermal energy, even considering the geysers in Iceland as an alternative to coal.

54 William J. Ashworth, 'Memory, Efficiency, and Symbolic Analysis: Charles Babbage, John Herschel, and the Industrial Mind', *Isis* 87, no. 4 (1996): 648.

3
The Machinery Question

Question: What is the effect of machinery?
Answer: To do that labour which must otherwise be done by hand, and to do it more perfectly and expeditiously.
Question: To whom then ought the machinery to belong?
Answer: To the men whose work it does – the labourers ...
Question: Who are the inventors of machinery?
Answer: Almost universally the working men.
Question: But why do not the working men use machinery for themselves?
No Answer!!!

The Pioneer, 1833[1]

The history of the English working classes begins in the second half of the eighteenth century with the invention of the steam engine and of machines for spinning and weaving cotton. It is well known that these incentives gave the impetus to the genesis of industrial revolution.

Friedrich Engels, *The Condition of the Working Class in England*, 1845[2]

---

1 *The Pioneer* 1, no. 4 (28 September 1833).
2 Friedrich Engels, *Die Lage der arbeitenden Klasse in England,* Leipzig: Wigand, 1845. Translated by W. O. Henderson and W. H. Chaloner as *The Condition of the Working Class in England*, Redwood City, CA: Stanford University Press, 1958, 9.

> *The science which compels the inanimate limbs of the machinery, by their construction, to act purposefully, as an automaton, does not exist in the worker's consciousness, but rather acts upon him through the machine as an alien power, as the power of the machine itself.*
>
> – Karl Marx, *Grundrisse*, 1858[3]

How to question technology

It is commonly believed that the Industrial Revolution took English society by storm and transformed an economy of regular agricultural cycles into one of violent and unstable growth. For a long time, across the political spectrum, it has been accepted that capitalism began as *machine capitalism* and that even the making and destiny of the working class was tied to an industrial machine.[4] Friedrich Engels himself declared this commonplace in the first line of *The Condition of the Working Class in England* (1845, quoted as an epigraph to this chapter). However, his collaboration with Karl Marx would radically challenge this apparently techno-deterministic view into one which recognised workers and the division of labour, rather than technology, as the main drivers of capitalist development.[5]

During the industrial age, machines came to supplant workers, dividing them into skilled and unskilled labourers, further separating mental from manual labour, and imposing new social hierarchies. Workers, however, resisted such division. They rebelled against machines and confronted their 'alien power', they discussed the role of machines, stormed factory floors to destroy them, and demanded, eventually, public education about them. The outcome of this outcry was the often-forgotten 'Machinery Question', the public debate sparked in English

---

3 Karl Marx, *Grundrisse: Foundations of the Critique of Political Economy*, London: Penguin 1973, 692–3.

4 The industrial age is today seen in a larger context and time scale. To clarify the debt of British industrial capitalism towards its colonies and limits of Marxism itself in regard to this aspect, see Cedric Robinson, *Black Marxism: The Making of the Black Radical Tradition*, Chapel Hill: University of North Carolina Press, 1983.

5 Engels's reading differs from Marx's in *Capital*, where the division of labour and the social relations of production are described as the main driver of technological development. Engels and Marx first met in 1844.

society at this time upon the massive replacement of workers by new technologies.

As a necessary prelude to the study of contemporary AI, this chapter aims to illuminate not just the social conflicts behind the Machinery Question but also the extended field of knowledge production around industrial machines and machine labour. Already during the industrial age, machines presented a problem of *machine intelligence* – also nderstood as lack of collective knowledge about them. Besides its well-celebrated muscular, energetic, and thermodynamic exploits (documented by Anson Rabinbach in his book *The Human Motor*),[6] industrial labour also entailed knowledge about machines, knowledge embodied by machines, and knowledge produced and projected anew by machines. This epistemic dimension of the industrial age, different from but related to its energetic one, is less investigated and appears always secondary in the vast literature of political economy (including Marxism). It is from this point of view – which is to say from the perspective of knowledge production, of the inquiry into the forms of knowledge of the industrial age – that this chapter hopes to cast a different light on and pay respect to a century of hard labour.

In her influential book *The Machinery Question and the Making of Political Economy*, historian Maxine Berg argues that at the time of the industrial age,

> machinery became the most immediate basis for the relationship between capitalist and worker. It was the machine which defined the organisation of work and which held the balance of power in the determination of the distribution of returns from labour.[7]

This is an egregious description of the field of forces, more precisely of the political battlefield, in which social and economic actors had to confront each other. Rather than a campaign of the industrialists, it must be immediately noted that the Machinery Question was first and

---

6 See Anson Rabinbach, *The Human Motor: Energy, Fatigue, and the Origins of Modernity*, Berkeley: University of California Press, 1992.

7 Maxine Berg, *The Machinery Question and the Making of Political Economy*, Cambridge: Cambridge University Press, 1980, 101.

foremost a reaction of the working class and an expression of its demand for control and ownership of technological progress. Berg writes:

> Workers criticised the rapid and unplanned introduction of new techniques in situations where the immediate result would be technological unemployment. But they also went beyond this to challenge the uses and property relations of technology. They demanded an equitable distribution of the gains from technical progress. Rather than raising the profits of the few, the machine, they argued, might lighten the labour and increase the leisure of the many. They also demanded greater control over the direction of technological change ... Technological progress should also be directed to changing the role of women in society, dispensing with the heavy manual labour and the household chores which prevented many women from claiming an equal position with men.[8]

The Machinery Question was canonically established by David Ricardo in the chapter 'On Machinery' that was added to the 1821 edition of his *Principles of Political Economy*. Ricardo's thesis was the following: while it was true that new machinery would cheapen commodity prices, nonetheless the working class would not benefit from this, since wages would be reduced by the competition among workers which is caused by technological unemployment. Berg adds that:

> The [machinery] question was central to everyday relations between master and workman, but it was also of major theoretical and ideological interest. The very technology at the basis of economy and society was a platform of challenge and struggle. The machine was debated at length in all sectors of society. It provoked the village cleric as much as it did the cosmopolitan intellectual; it concerned the politician as much as the workman and employer; the social reformer as much as scientist and inventor. These groups contended over the costs and benefits of the new technology. They hailed the release it provided from limits to growth, but disagreed over the impact it would have on wages, employment, and skill. They speculated on, and then either

---

8 Ibid., 17.

welcomed or dreaded, the changes the machine would bring to social relations. The origins and the ownership of machinery even came up for question. There was excitement and fear at this unknown force which swept relentlessly onward, casting the old society in its wake.[9]

The Machinery Question was therefore a complex phenomenon: an issue of popular culture, political propaganda, scientific contestation, and social control through education. The machine became the site of an intellectual struggle and political occupation by radical thinkers, utopian industrialists, and socialist (and sometimes conservative) militants.[10] The ideological struggle around machinery involved popular literature and pamphlets, poems and satires, and also the industrialists' celebration of a machine cult with dancing automata, 'mechanical Turks', and industrial engines set on display in public squares as tourist attractions. Charles Babbage was known for exhibiting a feminised dancing automaton in his salon, 'in the room next to the unfinished portion of the first Difference Engine'.[11] As Berg has stressed, the rise of political economy as a new discipline was part of the intellectual struggle to control the Machinery Question.[12]

The response to the employment of machines and workers' subsequent technological unemployment was also the demand, by both workers and industrialists, for more knowledge about machines, for more education and better training, which took the form of the Mechanics' Institute movement, among other initiatives. The year 1823 saw the establishment of the London Mechanics' Institute, later known as Birkbeck University – which still bears the Latin motto *In nocte consilium* ('Advice comes overnight'), as students were used (or, more accurately, forced) to attend evening and night courses after their daily shift at work. In 1826, Henry Brougham, future lord chancellor, founded the

---

9 Ibid., 2, 9–10.

10 Ibid., 168. This debate did not always take progressive directions. Thomas Carlyle was a racist thinker, for example, who often employed vitalist and gothic imageries against industrial machinery to attract workers to his cause. See Thomas Carlyle, 'Chartism', in *The Works of Thomas Carlyle*, ed. Henry Duff Traill, vol. 29, *Critical and Miscellaneous Essays IV*, Cambridge: Cambridge University Press, 2010, 118–204.

11 Simon Schaffer, 'Babbage's Dancer and the Impresarios of Mechanism', in *Cultural Babbage: Technology, Time and Invention*, ed. Francis Spufford and Jenny Uglow, London: Faber & Faber, 1996, 53–80.

12 Berg, *The Machinery Question*, 17.

Society for the Diffusion of Useful Knowledge to help those who had no access to schooling, the same year that the London University (later University College London) was founded. The Owenite Hawkes Smith went so far as to forge mechanical metaphors for education: 'There is an *intellectual* machinery, a *mental steam power* at work, and still rising in its action which renders education proportionately as cheap and as attainable to the man of small means as his clothing and his domestic appointments.'[13] It has indeed been forgotten how a good part of the British academic landscape finds its roots in the epistemic acceleration of the Industrial Revolution.[14]

Which *forms of knowledge* were debated under the auspices of the Machinery Question? Knowledge understood as skill and mechanical invention, applied rather than abstract science, was the mission of workers, engineers, educators, and industrialists. The Mechanics' Institutes were the incarnation of such a trend, searching, in Berg's words, for the 'optimal combination of science and skill . . . a higher form of skilled labour, one freed from the degradation of the division of labour and imbued with creative and innovative instincts'.[15] Everyone maintained that economic growth was bound to the invention of new machines and considered the *intellect as the machine within the machine*; however, the role of the intellect in this process was not then completely clear (and it is not yet today).

In this debate, Marx was probably one of the most original and acute voices. He came to question the technological determinism according to which the machine would be the prime mover of industrial capitalism. Reversing the common perception of the relation between technology and economy, he argued that technological development (the *means of production*) is triggered by the division of labour (the *relations of production*) and not the other way around.[16] According to Marx, capitalist accumulation is pushed by the exploitation of surplus value, by an

---

13 Hawkes Smith, 'On the Tendency and Prospects of Mechanics' Institutions', *The Analyst*, 1835.

14 For an account of the Mechanics' Institute movements as form of social control and ideological battleground, see Steven Shapin and Barry Barnes, 'Science, Nature and Control: Interpreting Mechanics' Institutes', *Social Studies of Science* 7, no. 1 (1977): 31–74.

15 Berg, *The Machinery Question*, 149.

16 Ibid., 158.

ever-growing division of labour, rather than by technological acceleration. In his view, it was the emerging intelligence of the division of labour, the spontaneous and distributed cognition of masters and workers, which invents machines, not science per se. Science came into play only afterwards, to improve machines that emerged from a social configuration. For Marx, to be precise, the actual *alien power* that moves capitalism is not machinery but living labour.

These sketches of the Machinery Question aim to reveal the full spectrum of social forces which lay behind the intellectual universe of the industrial age, rather than accept technological determinism as its main theory. Indeed, during that age a complex dialectics between social, technical, scientific, and cultural forms took place which cannot be reduced to any of them. The extended list of *knowledge models* and modalities of *knowledge production* that crossed the industrial age should at least include: the type of knowledge that is represented by the act of *invention* of a machine; the *division of labour* that inspires its *design*; the *know-how* of engineering and the *symbolic language* that is necessary to describe mechanism; *hard sciences* such as mechanics and thermodynamics; *non-technical disciplines* such as political economy; the *metrology* of manual and mental labour and the instruments to measure them; *collective knowledge* as embodied in both machinery and social relations (the so-called 'general intellect'); *educational movements* such as the Mechanics' Institutes; *political campaigns* such as the March of Intellect; and, finally, *popular mythologies* around automata such as the Mechanical Turk. The Machinery Question contained all these contested forms of knowledge. This was not entirely a public history – one of visible movements and effects – but more often an anonymous history: one of invisibilisation (of women's labour, especially) and of political amnesia surrounding early notions of political economy such as mental labour.

Against the spell of invisibilisation of the workforce, especially to better understand the demonisation of mental labour in the following century and the reason why mental labour disappeared from the debates on technology, this chapter expands, from opposite angles, upon the previous chapter's reflection on Babbage's labour theory of the machine. On the one hand, it illustrates the influence of knowledge in the definition of labour, framing a *knowledge theory of labour* (dear in the same way to nineteenth-century Ricardian socialists and twentieth-century knowledge economists). On the other hand, it illustrates the key influence of new machines and instruments on the development of new knowledge, expanding upon a *machine theory of science* that is key to the materialist epistemology of this book.

The knowledge theory of labour

The study of the relation between knowledge and labour has been complicated by the hegemony that science has maintained, in the modern era, in defining and enforcing social hierarchies. The epistemic imperialism of science institutions has obfuscated the role that labour, craftsmanship, experiments, and spontaneous forms of knowledge have played in technological change: it is still largely believed that only the application of science to industry can invent new technologies and prompt economic growth, while this is in fact rarely the case. Indeed, early nineteenth-century political economy already recognised the productive role of mental labour and the knowledge component of any form of manual labour in technological invention. Ricardian socialists such as William Thompson and Thomas Hodgskin, for instance, provided an analysis of mental labour that largely predates the theorists of the knowledge society of the twentieth century. Their position is illustrated, in this book, as a knowledge theory of labour, according to which the main component of labour is not muscular, physical, and energetic, but primarily psychological, intellectual, and informational.

William Thompson was 'an Irish landowner who embraced Owenism, and criticised political economy from a utopian socialist position, but on the basis of Ricardo's doctrines'.[17] In 1824, Thompson published a since-forgotten book with the optimistic title *An Inquiry into the Principles of the Distribution of Wealth Most Conducive to Human Happiness Applied to the Newly Proposed System of Voluntary Equality of Wealth*. There Thompson provided one of the first systematic definitions of knowledge labour of the modern age:

---

17 A footnote from Marx's *Grundrisse* introduces Thompson this way. Marx, *Grundrisse*, 537.

In speaking of labour, we have always included in that term the quantity of knowledge requisite for its direction. Without this knowledge, it would be no more than brute force directed to no useful purpose. In whatever proportion knowledge is possessed, whether in whole or in part, by the productive labourer, or by him who directs his labour, it is necessary in order to make his labour productive that some person should possess it.[18]

Presciently, Thompson argued that the economy of knowledge follows rules of diffusion that are different than the economy of scarcity of material goods and instead are driven by continuous expansion and free multiplication:

Wealth, the produce of labor, is necessarily limited in its supply . . . Not so with the pleasure derived from the acquisition, the possession, and diffusion of knowledge. The supply of knowledge is unlimited . . . The more it is diffused, the more it multiplies itself.

Thompson, however, perceived the ambivalence of instrumental knowledge, in a sort of 'dialectics of enlightenment' *ante litteram*. In a typical polemic of Owenism, Thompson described machinery as humiliating the 'general intellectual powers' of the workers that were reduced, in this way, to 'drilled automata'. The factory was an apparatus to keep the workers 'ignorant of the secret springs which regulated the machine and to repress the general powers of their minds' so 'that the fruits of their own labors were by a hundred contrivances taken away from them'.[19] Marx's quote from Thompson in *Capital* is a perfect distillation of this thought:

The man of knowledge and the productive labourer comes to be widely divided from each other, and knowledge, instead of remaining the handmaid of labour in the hand of the labourer to increase his productive powers . . . has almost everywhere arrayed itself against

---

18 William Thompson, *An Inquiry into the Principles of the Distribution of Wealth Most Conducive to Humane Happiness Applied to the Newly Proposed System of Voluntary Equality of Wealth*, London: Longman, Hurst, Rees, Orme, Brown and Green; Wheatley and Adlard, 1824, 272.

19 Thompson, *Principles of the Distribution of Wealth*, 272–5, 290, 292.

labour. 'Knowledge' becomes 'an instrument, capable of being detached from labour and opposed to it'.[20]

Thompson's was not only the first modern account of knowledge labour and of the cognitive component of all labour, but also one that recognised the alienation of knowledge from workers and its transformation into a repressive power inimical to the workers themselves.

Similar positions were advanced also by Thomas Hodgskin, a Ricardian socialist of libertarian tendency who believed in the progress of collective knowledge and the autonomy of society from both capital and state intervention. Hodgskin was one of the founders of the London Mechanics' Institute, where in 1826 he presented the lecture 'On the Influence of Knowledge', later published as part of his book *Popular Political Economy* (1827). Socialists such as Thompson and Hodgskin argued that knowledge is key to economic prosperity. Hodgskin complained that Adam Smith, the father of political economy, did not give a proper treatment to the subject, commenting:

> Those books, therefore, called Elements, Principles, or Systems of Political Economy, which do not embrace and fully develop ... the whole influence of knowledge on productive power, and do not explain the natural laws which regulate the progress of society in knowledge, are and must, as treatises on Political Economy, be essentially incomplete.[21]

In a clear anti-Malthusian argument, Hodgskin anchored the virtuous growth of knowledge to the needs of a growing population, in this way also reclaiming the territory of knowledge production from the monopoly of state academies and science institutions. He positively declared: 'Necessity is the mother of invention; and the continual existence of necessity can only be explained by the continual increase of people.'[22]

---

20 Karl Marx, *Capital: A Critique of Political Economy*, trans. Ben Fowkes, vol. 1, London; New York: Penguin Books; New Left Review, 1976, 482–483; Thompson, *Principles of the Distribution of Wealth*, 274.

21 Thomas Hodgskin, *Popular Political Economy: Four Lectures Delivered at the London Mechanics' Institution*, London: Tait, 1827, 97.

22 Ibid., 86. See Malthus's elitist account of knowledge in 'An Essay on the Principle of Population', 1798.

According to Hodgskin, it is the growth of population that demands better skill in producing and distributing wealth, thereby generating advanced knowledge: 'As the world grows older, and as men increase and multiply, there is a constant, natural, and necessary tendency to an increase in their knowledge, and consequently in their productive power.' Like Thompson, Hodgskin maintained that the rules of the knowledge economy are not those of capitalism: 'The laws which regulate the accumulation and employment of capital are quite dissimilar to and unconnected with the laws regulating the progress of knowledge.' In Hodgskin's view of society, there should be no intellectual hierarchies, no division of head and hand, no labour aristocracy to promote, because 'both mental and bodily labour are practised by almost every individual.'[23]

The demonisation of mental labour

As Berg noted, for participants in the Machinery Question and those within the Mechanics' Institute movement, the apparently benevolent celebration of craftsmanship was actually instrumental to dividing the working class and inciting a fabricated 'labour aristocracy' to mimic bourgeois customs:

> The rhetoric on the connection between technological progress and economic improvement in the Mechanics Institute Movement ... meant to contribute to the formalisation of hierarchies in the labour movement. The skilled artisan was to be separated from unskilled common labour, and both were to be detached from the middle class. This design for creating a 'labour aristocracy' was complemented by efforts to contribute to the discipline of the labour force.[24]

Cultivating the figure of the ingenious artisan among other gifted personalities (such as scientists and philosophers), the industrialist class aimed to divide the proletariat according to a hierarchy of deskilled and skilled workers, and to impose a gradual disciplining of labour. The workers' movement, on the other hand, fought to maintain a united

---

23 Ibid., 95, 78, 47.
24 Berg, *The Machinery Question*, 179.

front in which both unskilled workers and skilled artisans could perceive each other on the same side of the political confrontation. But in order to maintain such a position, for tactical reasons, it had to both conceal and absorb the difference of *mental labour* within the manual, and of individual labour within the collective. In terms of political strategy, in order to unify a divided front, it was therefore necessary to declare that all labour is manual (without implying that all labour is also mental). All collective knowledge, including skill, know-how, and even science, thus had to become an expression of *labour in common*.

Ultimately, this reaction against the social hierarchies of knowledge in the workers' movement led to the refusal of status for mental labour among its ranks and, in this way, the unconscious adoption of a bourgeois social segmentation. As such, it was in order to maintain the political unity of workers that mental labour was ostracised from the Machinery Question. The focus on manual labour has since then imposed an interpretation of labour as energetic performance only (as Rabinbach has noted, even Marx's *Arbeitskraft* was originally a notion from thermodynamics).[25] Both the middle class's discrimination of mental and manual labour and the working class's neutralisation of mental labour within manual labour were dictated by reasons of political tactics within the field of social forces of the industrial age. What is remarkable is that the twentieth-century amnesia surrounding the nineteenth-century theory of mental labour finds its explanation in the strength of the workers' movement in its confrontation with the capitalist class.

Marx had a specific role in organising the political amnesia of mental labour (see chapter 4). Although familiar with Thompson and Hodgskin, both of whom he quoted, in *Capital* Marx removed all references to mental labour, knowledge labour, and the 'general intellect' to replace them with the inventive capacity of the division of labour and the new figure of the collective worker, or *Gesamtarbeiter*. Following Babbage, Marx adopted the idea that the extended division of labour, rather than science, was the inventor of the machine. In this way, Marx reversed Thompson and Hodgskin's *knowledge theory of labour* into the more materialistic *labour theory of knowledge*, in which forms of labour that are spontaneous, unconscious, tacit, and collective are also eventually recognised as producing knowledge. Industrial machinery, nonetheless,

---

25 See Rabinbach, *The Human Motor*.

The Machinery Question

ended up polarising the distance between skilled and deskilled labour. As Marx sharply summarised: 'By the introduction of machinery the division of labour inside society has grown up, the task of the worker inside the workshop has been simplified, capital has been concentrated, human beings have been further dismembered.'[26] It is on this basis which Berg concludes that 'machinery did not displace labour. Rather, it differentiated this labour by dismembering the old craft.'[27]

The distinction of head and hand, of mental and manual labour, is not only typical of modern industrial societies; it has been part of Western culture at least since the Aristotelian opposition of *episteme* ('knowledge') and *techne* ('art' or 'craft') in ancient Greece, which later became functional for defining social hierarchies across the West. Historians of mathematics such as Peter Damerow actually predate the social separation of mental and manual labour to the dawn of civilisations due to the need to count populations, plan agriculture, and administer resources. The control of abstract symbols would later develop into the domain of letters and spirit and a long-lasting class segmentation of society, as historians of science Lissa Roberts and Simon Schaffer record:

> Self-appointed mental workers, such as philosophers, scientists, policy-makers and bureaucrats, then as now, claimed and constructed the dominion of their 'understanding' over hand-workers and their crafts. They relied on the mutual reinforcement of coercive rhetoric and brutal deed. The easy acceptance of their categories has left us with a historical map shaped by oppositional and hierarchically ordered pairs: scholar / artisan, science / technology, pure / applied and theory / practice.[28]

It could therefore be argued that the first *division of labour* in the modern sense is the separation of head from hand that gradually emerged out of the workshops of the Renaissance to be fully severed

---

26 Karl Marx, *The Poverty of Philosophy. Answer to the 'Philosophy of Poverty' by M. Proudhon*, Moscow: Progress Publishers, 1955.
27 Berg, *The Machinery Question*, 34.
28 Lissa Roberts, Simon Schaffer, and Peter Dear (eds), *The Mindful Hand: Inquiry and Invention from the Late Renaissance to Early Industrialisation*, History of Science and Scholarship in the Netherlands, vol. 9, Amsterdam: Koninkliijke Nederlandse Akademie van Wetenschappen, 2007, xiii.

precisely in the industrial factories as the division of mental and manual labour. For instance, Edgar Zilsel, a historian of science, documented how even the 'heroes' of the so-called Scientific Revolution, such as Galileo Galilei, learned more in clandestine workshops, hidden libraries, and nomadic classrooms than at universities.[29] Roberts and Schaffer, for their part, have proposed the elegant image of the 'mindful hand' as a way to recognize and recompose the ingenuity of manual labour, mechanical experiments, and scientific workshops throughout modernity, without romanticising craftsmanship as conservative discourse so often does.[30] Rather than cultivating the provincial 'heroism' of craftsmen in a reactionary way, the image of the 'mindful hand' stresses the convivial dimension of experimental life and its inventions.

Industrial modernity has established itself on the capture of this collective knowledge by state and economic apparatuses, by institutions of knowledge and technologies of knowledge, which eventually turned mental labour into a *Geist*, to use the ambivalent German term – a ghost, more than an intellectual spirit, that political theory still struggles to grasp. Extorted from workers and social cooperation, mental labour assumed the nature of a half-visible demon: a political issue to be exorcised, for opposite reasons, by the workers' movement as much as by corporate interests.[31]

The machine theory of science

Tool-makers and machine operators knew that they were contributing to the invention of new technologies. What they were rarely aware of is that they were also contributing to new scientific discoveries. New machines prompt scientific notions and paradigm shifts more often than science happens to invent new technologies from above. As in an

---

29 Edgar Zilsel, *The Social Origins of Modern Science*, History of Philosophy of Science, Dordrecht: Springer, 2002, 5.

30 For a progressive reading of craftmanship, see Richard Sennet, *The Craftsman*, New Haven, CT: Yale University Press, 2008. For a conservative reading of craftsmanship, see Peter Sloterdijk, *You Must Change Your Life*, London: John Wiley & Sons, 2014, 292: 'Whoever has no interest in craftsmen should therefore be equally silent about heroes.'

31 Roberts and Schaffer address this ambivalence in the idea of the modern 'denial of cunning'.

example mentioned earlier, it was the steam engine which gave birth to thermodynamics, rather than the other way around. The science of heat and energy transformation developed to ameliorate the steam engine: it was a projection of the lucrative ambitions of autonomous motion, not just the child of curiosity towards the universe. In the study of the forms of knowledge that undergo mechanisation, it is important to also highlight the knowledge of the world that is expressed anew by machines. The idea that tools, instruments, and machines project and constitute the ontology of scientific theories about the world can be defined as a *machine theory of science*. As Peter Damerow and Wolfgang Lefèvre have stressed, among others, the tools of work are also tools for exploring the world and speculating upon it: 'The development of science depends on the development of its material tools . . . When using a material tool, more can always be learned than the knowledge invested in its invention.'[32] Yet Damerow and Lefèvre also stress that science is never fully independent from the materiality of its instruments:

> Science is not free in forming its abstractions; in this activity it is restricted by material preconditions, more precisely, by the specific tools at its disposal that provide cognition with abstractions which are capable of realization . . . The material tools of scientific labor define a scope of objective possibilities that represent the framework for developing scientific abstractions.[33]

Tools and machines, however, are never fully transparent in their implications. Machines are born as experiments, and they are often operated without full knowledge of their workings. Science is developed to cover these blind spots in our knowledge of machines, not just of the universe. On the other hand, the perception of nature is often machine based, not simply because of the mediation of instruments on perception but because machines have influenced, indirectly, the ontology of entire scientific paradigms. For example, into the twenty-first century,

---

32 Peter Damerow and Wolfgang Lefèvre, 'Tools of Science', in Peter Damerow, *Abstraction and Representation: Essays on the Cultural Evolution of Thinking*, Berlin: Springer, 2013, 401.
33 Ibid., 400.

the standard theory of time remains based on the irreversible arrow of entropy that was encountered and conceptualised, for the first time, in the chambers of the steam engine before being canonised in the second principle of thermodynamics.[34] It is not an exaggeration to say that the universe is still perceived nowawdays as from the belly of an industrial engine.

In 1931, with his lecture 'The Social and Economic Roots of Newton's *Principia*', Boris Hessen consolidated the history of science and technology upon the method of historical materialism, arguing that Newton's endeavours were indebted to the tools and machines of the time which happened to be, unsurprisingly, also the main *means of production* of the time: water canal transport techniques, water pumps and pulleys used in the mining industry, new firearms and their ballistic control, and so forth.[35] More recently, Peter Galison has accounted for the difference between Newtonian and Einsteinian physics through analysing the technical perception of time in their respective ages and the definition of synchronisation: a centralised universal clockwork maintained a uniform time in Newton's case, whereas it was 'an electromechanical world' connected by new networks of communication such as the telephone in Einstein's case.[36] Going back to the industrial age, the historical epistemology of science and technology suggests that we reconsider the project of machine intelligence as a prism reflecting multiple forms of knowledge. Stretching its definition across a larger time scale, the expression 'machine intelligence' ultimately acquires at least four meanings: (1) the human knowledge of the machine; (2) the knowledge embodied by the machine's design; (3) the human tasks automated by

---

34 Arthur Stanley Eddington, *The Nature of the Physical World*, Cambridge: Cambridge University Press, 1928; Harold Francis Blum, *Time's Arrow and Evolution*, Princeton, NJ: Princeton University Press, 1955.

35 Now included in Gideon Freudenthal and Peter McLaughlin (eds), *The Social and Economic Roots of the Scientific Revolution: Texts by Boris Hessen and Henryk Grossmann*, Berlin: Springer, 2009.

36 The historical epistemology of science can take different names: Andrew Pickering has called it a 'cyborg history', Galison a 'technological reading' of science's theoretical developments, Henning Schmidgen a 'machine history' (*Maschinen-Geschichte*) of science. Andrew Pickering, 'Cyborg History and the World War II Regime', *Perspectives on Science* 3, no. 1 (1995): 1–48; Peter Galison, *Einstein's Clocks, Poincaré's Maps: Empires of Time*, New York: Norton, 2003, 290; Henning Schmidgen, *Hirn und Zeit: Die Geschichte eines Experiments, 1800–1950*, Berlin: Matthes & Seitz 2014, 44.

the machine; and (4) the new knowledge of the universe made possible by its use.

The Machinery Question in the age of AI

The industrial machine is a powerful artefact because it imbricates in one thing the relations between energy and matter, knowledge and science, but more importantly between capital and labour. In this sense, the industrial machine appears to be the incarnation of the many contradictions of capitalism and a concrete locus of a social and ideological struggle. A similar fascination with the political centrality of technology has extended until the present day, reiterating ambivalent impressions of the industrial age. Both academic techno-determinism and corporate techno-solutionism, for instance, consider it today as the core of the political question. However, it would be a gross mistake to consider technology the *unique locus* of political conflict. As this book is trying to explicate, social relations and in particular labour cooperation are the 'engines' of technical and political development. But, in their own terms, such social relations and the category of labour itself have to be scrutinised. What is labour cooperation made of, by the way? How was the notion of labour constructed, employed, narrated, and analysed by the political economy of the nineteenth and twentieth century? This is not a trivial question, because the idea of labour that is still used today is an inheritance from the nineteenth century and a product of the bold political confrontations of that time: of labour as a manual activity often devoid of any mental component.

Industrial capitalism was not only an energetic intensification of labour and production; it was also a transformation of the division of labour and social relations, to the point of becoming the matrix of a new kind of knowledge production – not only mathematics, mechanics, and physics, but knowledge of the most diverse kinds. In the early nineteenth century, Ricardian socialists such as Thompson and Hodgskin were already discussing the social potentialities and psychic implications of 'mental labour' claiming that knowledge is the first source of labour. Other political economists, such as Marx, agreed but argued that both mental and manual labour, without distinction, were the source of collective knowledge. It was close to the workshops of the industrial age that modern computation, eventually, was born as the project to mechanise the division of mental labour, as Babbage experimented with his calculating engines.

This chapter has explored the hypothesis that during the industrial age, knowledge and intelligence comprised the true hidden transaction between labour and capital. As we have seen, all labour, without distinction, was and still is cognitive and knowledge-producing. The most important component of labour is not energy and motion (which are easy to automate and replace) but knowledge and intelligence (which are far from being completely automated in the age of AI). The industrial age was also the moment of the *originary accumulation of technical intelligence* as the dispossession of knowledge from labour. AI is today the continuation of the same process: it is a systematic mechanisation and capitalisation of collective knowledge into new apparatuses, into the datasets, algorithms, and statistical models of machine learning, among other techniques. Ultimately, it is not difficult to imagine AI as a late avatar of the collective worker, the *Gesamtarbeiter* that was for Marx the main actor of industrial production. As we shall see, the nineteenth-century Machinery Question is also of signal importance for figuring out how to question this generalised process of automation in the age of AI. Aptly, in 2016 the *Economist* issued a special report on AI forewarning 'The Return of the Machinery Question'.[37]

---

37 'The Return of the Machinery Question', special report, *Economist*, June 2016, economist.com.

4

The Origins of Marx's General Intellect

> *The general intellect of the whole community, male and female, is stunted or perverted in infancy, or more commonly both, by keeping from women the knowledge possessed by men . . . The only and the simple remedy for the evils arising from these almost universal institutions of the domestic slavery of one half the human race, is utterly to eradicate them. Give men and women equal civil and political rights.*
>
> William Thompson, *An Inquiry into the Principles of the Distribution of Wealth*, 1824[1]

> *It is nearly twenty years since the first impulse was given to the general intellect of this country, by the introduction of a new mechanical system for teaching reading and writing, by cheaper and more efficacious methods than those previously in use . . . The public mind has infinitely advanced: in despite of all the sneers at the phrase of the 'march of intellect', the fact is undeniable, that the general intellect of the country has greatly progressed. And one of the first fruits of extended intelligence has been the conviction, now fast becoming*

---

1 William Thompson, *An Inquiry into the Principles of the Distribution of Wealth Most Conducive to Human Happiness Applied to the Newly Proposed System of Voluntary Equality of Wealth*, London: Longman, 1824, 214.

universal, that our system of law, so far from being the best in the world, is an exceedingly bad one; and stands in the most pressing need of revision and reform.

London Magazine, 1828²

The development of fixed capital indicates to what degree general social knowledge has become a direct force of production, and to what degree, hence, the conditions of the process of social life itself have come under the control of the general intellect and been transformed in accordance with it.

Karl Marx, Grundrisse, 1858³

The March of Intellect

An 1828 caricature by cartoonist William Heath from the series 'March of Intellect' depicts a giant automaton advancing with long strides and holding a broom to sweep away a dusty mass of clerks, clergy, and bureaucrats, representing figures of the old order and obsolete laws (see fig. 4.1). The automaton's belly is a steam engine, while its head is made of books of history, philosophy, and (importantly) mechanics. Its crown reads 'London University'. In the background, the goddess of justice lies in ruins, summoning the automaton: 'Oh come and deliver me!!!' While at first the cartoon might seem a paean to democratic ideals and intellectual advance, on closer observation, the caricature is intended to ridicule the belief that the technologies of industrial automation (already resembling robots) might become a true agent of political change and social emancipation under the command of public education. Indeed, Heath's series of satirical engravings was originally commissioned by the Tories to voice their sarcasm regarding a potential democratisation of knowledge and technology across all classes. Nonetheless, by dint of his visionary pen, they became an accidental manifesto for the progressive camp and the invention of the future.[4]

---

2 'Education of the People', *London Magazine*, April 1828, 1; 'Reforms in the Law, No. I. The History of a Suit', *London Magazine*, June 1828, 309.

3 Karl Marx, *Grundrisse: Foundations of the Critique of Political Economy*, trans. Martin Nicolaus, London: Penguin, 1993, 706.

4 'Even though Heath was satirising the movement, his posters include some wonderful future ideas for transport, including a steam horse and a steam coach, a

The Origins of Marx's General Intellect

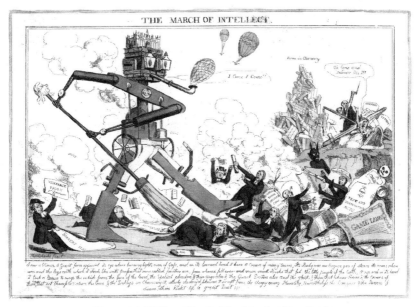

Figure 4.1. William Heath, 'The March of Intellect', ca. 1828, print, British Museum.

Initiated as a campaign in England during the Industrial Revolution, the March of Intellect, or 'March of Mind', demanded the amelioration of society's ills through programmes of public education for the lower classes.[5] The expression 'March of Intellect' was introduced by the industrialist and utopian socialist Robert Owen in a letter to the *Times* in 1824, remarking that in recent years 'the human mind has made the most rapid and extensive strides in the knowledge of human nature, and in general knowledge'.[6] The campaign triggered a reactionary and unsurprisingly racist backlash: the *Times* started to mock the ambitions of the working class under sarcastic headlines of the worst colonial mentality such as 'The March of Intellect in Africa'.[7] As a campaign for progress in

---

vacuum tube, a bridge to Cape Town, and various forms of flight, including a flying postman.' Mike Ashley, 'Inventing the Future', British Library blog, 15 May 2014, bl.uk.

5 See Don Herzog, *Poisoning the Minds of the Lower Order*, Princeton, NJ: Princeton University Press, 2000.

6 Ashley, 'Inventing the Future'.

7 See Michael Hancher, 'Penny Magazine: March of Intellect in the Butchering Line', in *Nineteenth-Century Media and the Construction of Identities*, Laurel Brake, Bill Bell and David Finkelstein, London: Palgrave, 2016, 93.

both literacy and technology, the March of Intellect was part of the so-called Machinery Question examined in the previous chapter. In 1828 the *London Magazine* endorsed the March of Intellect for the benefit of the 'general intellect of the country' – which, its editors argued, thanks to mass education, would understand the need to reform a decaying legislative system.[8] When in 1858 Marx used the expression (in English) 'general intellect' in the famous 'Fragment on Machines' of the *Grundrisse*, he was echoing the political climate of the March of Intellect and the power of 'general social knowledge' to, in his reading, weaken and subvert the chains of capitalism rather than those of old institutions.[9]

But it was specifically in William Thompson's *An Inquiry into the Principles of the Distribution of Wealth* (published in 1824, the same year that Owen launched the March of Intellect) that Marx first encountered the idea of the general intellect and, more importantly, the argument that knowledge, once it has been alienated by machines, may become a power inimical to workers.[10] The book contains what is probably the first systematic account of mental labour – followed by Thomas Hodgskin's account in *Popular Political Economy* (1827) and Charles Babbage's project to mechanise mental labour in *On the Economy of Machinery and Manufactures* (1832).[11] Afterwards, because of the decline of the Mechanics' Institutes and tactical decisions within the workers' movement, the notion of mental labour encountered a hostile destiny in the Machinery Question.

Given this backdrop, when twentieth-century authors began to analyse the so-called knowledge society and thought they were discussing for the first time forms of symbolic, informational, and digital labour, they were actually operating in an area of political amnesia. In fact, Marx himself was partly responsible for bringing about this amnesia.[12] While he engaged with Thompson's and

---

8 *London Magazine*, April and June 1828 issues. See chapter epigraph.
9 Karl Marx, *Grundrisse: Foundations of the Critique of Political Economy*, trans. Martin Nicolaus, London: Penguin, 1993, 690–71.
10 Thompson, *Principles of the Distribution of Wealth*.
11 Thomas Hodgskin, *Popular Political Economy: Four Lectures Delivered at the London Mechanics Institution*, London: Tait, 1827; Charles Babbage, *On the Economy of Machinery and Manufactures*, London: Charles Knight, 1832.
12 In *Capital* (493) Marx refers to Wilhelm Schulz's distinction between tool and machine, yet without commenting on Schulz's account of intellectual production

Hodgskin's political economy, he considered their emphasis on mental labour as the celebration of individual creativity – as the cult of the gifted artisan, the ingenious tool-maker, and the brave engineer – *against* labour in common: in *Capital*, Marx intentionally replaced the mental labourer with the 'collective worker' or *Gesamtarbeiter*. Marx's refusal to employ the concept of mental labour was due to the difficulty of mobilising collective knowledge into campaigns on the side of workers. The substance of knowledge and education is such that they can only be summoned for universalist battles (for the 'general intellect of the country') rather than partisan ones on the side the proletariat. Besides, since *The German Ideology*, Hegel's notion of absolute spirit appeared to be the antagonist of Marx's method of historical materialism: Marx transposed his famous anti-Hegelian passage 'life is not determined by consciousness, but consciousness by life' to industrial England, in order to claim that labour is not determined by knowledge, but knowledge by labour.[13]

Traditionally, for Marxism, the distinction between manual and mental labour evaporates in the face of capital insofar as any kind of labour is *abstract labour* – that is, labour measured and monetised for the benefit of producing surplus value. What follows shares this traditional starting point but goes on to depart from orthodox Marxist positions. I wish to consider that any machinic interface with labour is a social relation, as much as capital, and that the machine, as much as money, mediates the relation between labour and capital.[14] Thinking with, as well as beyond, Marx, I want to stress that any technology influences the metrics of abstract labour. For this purpose, this chapter traces

---

(*geistige Produktion*) from *Die Bewegung der Produktion* (Zurich, 1843). Karl Marx, *Das Kapital. Kritik der politischen Ökonomie*, vol. 1, Hamburg: Meissner, 1867; *Capital: A Critique of Political Economy*, vol. 1, trans. Ben Fowkes, London: Penguin, 1981. See Walter Grab, *Dr. Wilhelm Schulz aus Darmstadt. Weggefährte von Georg Büchner und Inspirator von Karl Marx*, Frankfurt am Main: Gutenberg, 1987. Thanks to Henning Schmidgen for this reference.

13 Karl Marx and Friedrich Engels, *Die deutsche Ideologie* (1846), 1st ed., Moscow: Marx-Engels Institute, 1932; Karl Marx and Friedrich Engels, *The German Ideology*, ed. C. J. Arthur, New York: International Publishers, 1970.

14 This idea could be termed a *labour theory of value mediated by machinery*, It points to a further problematic in which technology would take over some features of the money form.

the origins of Marx's general intellect in order to reconsider unresolved issues of early political economy, such as the econometrics of knowledge, that are increasingly relevant today.[15] In the current debates on the alienation of collective knowledge into corporate AI, we are, in fact, still hearing the clunky echoes of the nineteenth-century Machinery Question.

The discovery of Marx's 'Fragment on Machines'

Sophisticated notions of mental labour and the knowledge economy were offered at the dawn of the Victorian age, and already then were given radical interpretations. Marx, for example, addressed the economic roles of skill, knowledge, and science in his *Grundrisse*, specifically in the section that has become known as the 'Fragment on Machines'. There Marx explored an unorthodox hypothesis which was not to be reiterated in *Capital*: that because of the accumulation of the general intellect (particularly as scientific and technical knowledge embodied in machinery), labour would become secondary to capitalist accumulation, causing a crisis for the labour theory of value and blowing the foundations of capitalism skywards.[16] After 1989, Marx's 'Fragment on Machines' was revived by Italian *post-operaismo* as a prescient critique of the transition to post-Fordism and the paradigms of a knowledge society and an information economy.[17] Since then,

---

15 In the nineteenth century, physiologists and political economists tried to figure out a 'metrology' of 'cerebral labour'; according to Schaffer, the attempts to quantify intelligence with the aid of instruments contributed to the project of artificial intelligence in the following century. See Simon Schaffer, 'OK Computer', in *Ecce Cortex: Beitraege zur Geschichte des modernen Gehirns*, ed. Michael Hagner, Göttingen: Wallstein Verlag, 1999, 254–85.

16 This visionary hypothesis did not emerge again in *Capital*, again as a result of historical circumstances. Notebooks 6 and 7 were written in the winter of 1857–58, amid a financial crisis, whereas *Capital* was published after the crisis was over.

17 The primary sources of the complex debate on Marx's general intellect can be succinctly reconstructed as follows: Paolo Virno, 'Citazioni di fronte al pericolo', *Luogo comune* 1 (November 1990), translated by Cesare Casarino as 'Notes on the General Intellect', in *Marxism beyond Marxism*, ed. Saree Makdisi et al., New York: Routledge, 1996, 265–72; Christian Marazzi, *Il posto dei calzini: La svolta linguistica dell'economia e i suoi effetti sulla politica*, Torino: Bollati Boringhieri, 1994, translated by Giuseppina Mecchia as *Capital and Affects: The Politics of the Language Economy*, New York:

many authors – including some outside Marxism – have mobilised this esoteric fragment as a prophecy of different economic crises, especially following the internet bubble and 2000 stock market crash. The way Marx's 'Fragment on Machines' has reached even the debate on artificial intelligence and post-capitalism is a philological adventure that is worth recapitulating.[18]

The *Grundrisse* is 'a series of seven notebooks rough-drafted by Marx, chiefly with the purpose of self-clarification, during the winter of 1857–8'.[19] Indeed, the notebooks frequently reveal the method of inquiry and subtext of *Capital*, published a decade later. Yet the *Grundrisse* remained unpublished until the twentieth century – in Moscow in 1939 and Berlin in 1953 – which means that its reception entered Marxist debates almost a century after the publication of *Capital*. While a partial Italian translation started to circulate in 1956, a complete English translation was to become available only in 1973.[20] The denomination 'Fragment on Machines' to define specifically notebooks 6 and 7 of the *Grundrisse* became canonical due to the editorial choice of Raniero Panzieri, who published their translation under the title 'Frammento sulle macchine' in the 1964 issue of *Quaderni Rossi*, the journal of Italian *operaismo*.[21] In

---

Semiotext(e), 2011; Maurizio Lazzarato and Antonio Negri, 'Travail immatériel et subjectivité', *Futur antérieur* 6 (1991): 86–99; Paolo Virno, *Grammatica della Moltitudine*, Rome: Derive Approdi, 2002, translated by Isabella Bertoletti et al. as *A Grammar of the Multitude*, New York: Semiotext(e), 2004; Carlo Vercellone, 'From Formal Subsumption to General Intellect: Elements for a Marxist Reading of the Thesis of Cognitive Capitalism', *Historical Materialism* 15, no. 1 (2007): 13–36. Probably the first reception of this debate in English is Nick Dyer-Witheford, *Cyber-Marx: Cycles and Circuits of Struggle in High-Technology Capitalism*, Champaign: University of Illinois Press, 1999. See Michael Hardt and Antonio Negri, *Empire*, Cambridge, MA: Harvard University Press, 2000. For a critique of *operaismo*'s interpretation, see Michael Heinrich, 'The Fragment on Machines: A Marxian Misconception in the *Grundrisse* and Its Overcoming in *Capital*', and Tony Smith, 'The General Intellect in the *Grundrisse* and Beyond', In *Marx's Laboratory: Critical Interpretations of the Grundrisse*, ed. Riccardo Bellofiore et al., Leiden: Brill, 2013, 195–212, 213–31.

18 See Paul Mason, 'The End of Capitalism Has Begun', *Guardian*, 17 July 2015; and Paul Mason, *Postcapitalism: A Guide to Our Future*, London: Macmillan, 2016. See also MacKenzie Wark, *General Intellects*, London: Verso, 2017.

19 Martin Nicolaus, 'Foreword', in Marx, *Grundrisse*, 7.

20 Marcello Musto, 'Dissemination and Reception of the Grundrisse in the World', in *Karl Marx's Grundrisse*, ed. Marcello Musto, London: Routledge, 2008, 207–16.

21 Karl Marx, 'Frammento sulle macchine', trans. Renato Solmi, *Quaderni Rossi* 4 (1964).

the same year, the German philosopher Herbert Marcuse drew upon notebooks 6 and 7 in his *One-Dimensional Man*, while discussing the emancipatory potential of automation.[22] In 1972, in a footnote in *Anti-Oedipus*, Gilles Deleuze and Felix Guattari also refer to them as the 'chapter on automation'.[23] That same year, they were partially published in English as 'Notes on Machines' in the journal *Economy and Society*.[24] In 1978 Antonio Negri gave an extended commentary on the 'chapter on machines' in his *Marx Beyond Marx* seminar in Paris (at the invitation of Louis Althusser), reading it against the background of the social antagonism of the preceding decade. But it was only after the fall of the Berlin Wall that Italian *post-operaismo* rediscovered and promoted the 'Fragment on Machines'. In 1990 the philosopher Paolo Virno drew attention to the notion of general intellect in the journal *Luogo comune*. Paying ironic tribute to the Spaghetti Western, he was already warning about the cycles of the concept's revival:

> Often in westerns the hero, when faced by the most concrete of dilemmas, cites a passage from the Old Testament ... This is how Karl Marx's 'Fragment on machines' has been read and cited from the early 1960s onwards. We have referred back many times to these pages ... in order to make some sense out of the unprecedented quality of workers' strikes, of the introduction of robots into the assembly lines and computers into the offices, and of certain kinds of youth behavior. The history of the 'Fragment's' successive interpretations is a history of crises and of new beginnings.[25]

Virno explained that the 'Fragment' was quoted in the 1960s to question the supposed neutrality of science in industrial production, in the 1970s as a critique of the ideology of labour in state socialism, and, finally, in the 1980s as a recognition of the tendencies of post-Fordism, yet without any emancipatory or conflictual reversal, as Marx would

---

22 Herbert Marcuse, *One-Dimensional Man*, Boston: Beacon Press, 1964, 39.
23 Gilles Deleuze and Félix Guattari, *Anti-Oedipus: Capitalism and Schizophrenia*, vol. 1, trans. Robert Hurley et al., Minneapolis: University of Minnesota Press, 1983, 232n76. See also Matteo Pasquinelli, 'Italian Operaismo and the Information Machine', *Theory, Culture and Society* 32, no. 3 (2015).
24 Karl Marx, 'Notes on Machines', *Economy and Society* 1, no. 3 (1972).
25 Virno, 'Notes'.

have wished. While Marxist scholars aimed for greater philological rigour in their reading of the general intellect, militants updated its interpretation in the context of current social transformations and struggles.[26] *Post-operaismo* famously forged new antagonistic concepts out of Marx's general intellect, such as 'immaterial labour', 'mass intellectuality', and 'cognitive capitalism', stressing the autonomy of 'living knowledge' against capital. A lesson worth recalling from the Machinery Question discussed in the previous chapter, however, is that the issue of collective knowledge should never be separated from its embodiment in machines, instruments of measurement, and *Kulturtechniken*. Indeed, the employment of artificial intelligence in the twentieth century has abruptly reminded everyone that knowledge can be analysed, measured, and automated as successfully as manual labour.

Scholars have wondered where the expression 'general intellect' came from, as it appears only once, in English, in the *Grundrisse*. Virno thought he detected the echo of Aristotle's *nous poietikos* and Rousseau's *volonté générale*.[27] As the 'Fragment' follows strains of argumentation that are similar to chapters 14 and 15 of *Capital* on the division of labour and machinery, it is not surprising that the missing sources can be found in the footnotes to these chapters of *Capital*. These common strains of argumentation echo, fundamentally, Babbage's theory of machinery, and it is by following Marx's reading of Babbage in chapter 14 of *Capital* that the notion of general intellect can be reliably traced back to William Thompson's notion of 'knowledge labour'.

Marx's interpretation of Babbage

In 1832, Babbage advised his fellow industrialists, 'The workshops of [England] contain within them a rich mine of knowledge, too generally

---

26 Wolfgang Fritz Haug warned that the nebulous origins of the general intellect contributed to a sloganistic use 'at the cost of theoretical arbitrariness'. The general intellect belongs, Haug asserts, to a galaxy of similar Marxian terms to be taken in consideration, such as 'general social labour', 'general scientific labour', 'accumulation of knowledge and of skill, the general productive forces of the human brain', 'general progress', 'development of the general powers of the human head', 'general social knowledge', 'social intellect'. Wolfgang Fritz Haug, 'Historical-Critical Dictionary of Marxism: General Intellect', *Historical Materialism* 18, no. 2 (2010).

27 Virno, 'Notes'.

neglected by the wealthier classes.'[28] Following this invitation to the industrial workshops as 'mundane places of intelligence', Simon Schaffer finds that 'Babbage's most penetrating London reader' was Marx.[29] Indeed, Marx had already quoted Babbage in *The Poverty of Philosophy* during his exile in Brussels in 1847 and, since then, adopted two analytical principles that were to become pivotal in *Capital* in drawing a robust theory of the machine and in grounding the theory of relative surplus value.

The first is the *labour theory of the machine*, which states that a new machine comes to imitate and replace a previous division of labour. As examined previously, this is an idea already formulated by Adam Smith, but better articulated by Babbage due to his greater technical experience. The second analytical principle is the 'Babbage principle', also discussed earlier, which has been renamed here the *principle of labour calculation*. It states that the organisation of a production process in small tasks (division of labour) allows exactly the necessary quantity of labour to be purchased for each task (division of value). In this respect, the division of labour provides not only the design of machinery but also an economic configuration to calibrate and calculate surplus labour extraction. In complex forms of management such as Taylorism, the principle of surplus labour modulation opens onto a clockwork view of labour, which can be further subdivided and recomposed into algorithmic assemblages. The synthesis of both analytical principles ideally describes the machine as an apparatus that actively projects back a new articulation and metrics of labour. In the pages of *Capital*, the industrial machine appears to be not just a regulator to discipline labour but also a calculator to measure relative surplus value, echoing the numerical exactitude of Babbage's calculating engines.

Here, I will read the *Grundrisse* and *Capital* through the lens of Babbage's two analytical principles. We will see how Babbage's labour theory of the machine is used by Marx to raise the figure of the collective worker as a sort of reincarnation of the general intellect and, furthermore, how Babbage's principle of modulation of surplus labour is used to sketch the idea of relative surplus value. Taken together, Babbage's

---

28 Babbage, *On the Economy of Machinery*, vi.
29 Simon Schaffer, 'Babbage's Intelligence: Calculating Engines and the Factory System', *Critical inquiry* 21, no. 1 (1994), 204.

two principles show that the general intellect of the *Grundrisse* evolves in *Capital* into a machinic collective worker, almost with the features of a proto-cybernetic organism, and the industrial machine becomes a calculator of the relative surplus value that this cyborg produces.

In discussing the relation between labour and machinery, knowledge and capital, Marx found himself embedded in a hybrid dialectics between German idealism and British political economy. The similar argumentation in the *Grundrisse* and *Capital* in the sections on machinery and division of labour follows four movements, to which I will now turn: (1) the invention of machinery through the division of labour, (2) the alienation of knowledge by machinery, (3) the devaluation of capital by knowledge accumulation, and (4) the rise of the collective worker.

The invention of machinery through the division of labour

Who is the inventor of the machine? The worker, the engineer, or the factory's master? Science, cunning, or labour? As a fellow of the Royal Society, Babbage publicly praised the gifts of science, but theoretically maintained that machinery emerges as a replacement of the division of labour. As already discussed, Babbage was committed to a *labour theory of the machine*, since, for him, the design of a new machine always imitates the design of a previous division of labour. In *The Poverty of Philosophy* (1847), Marx already mobilised Babbage against Proudhon, who thought that machinery is the *antithesis* of the division of labour. Marx argued the opposite, that machinery emerges as the *synthesis of the division of labour*: 'When, by the division of labour, each particular operation has been simplified to the use of a single instrument, the linking up of all these instruments, set in motion by a single engine, constitutes – a machine.'[30] Later, in the *Grundrisse*, Marx kept on drawing on Babbage to remark that technology is not created by the 'analysis' of nature by science but by the 'analysis' of labour:

> It is, firstly, the analysis [*Analyse*] and application of mechanical and chemical laws, arising directly out of science, which enables

---

30 Babbage, *On the Economy of Machinery*, as quoted in Karl Marx, *The Poverty of Philosophy*, Moscow: Progress Publishers, 1955, 121.

the machine to perform the same labour as that previously performed by the worker. However, the development of machinery along this path occurs only when large industry has already reached a higher stage, and all the sciences have been pressed into the service of capital . . . Invention then becomes a business, and the application of science to direct production itself becomes a prospect which determines and solicits it. But this is not the road along which machinery, by and large, arose, and even less the road on which it progresses in detail. This road is, rather, dissection [*Analyse*] – through the division of labour, which gradually transforms the workers' operations into more and more mechanical ones, so that at a certain point a mechanism can step into their places.[31]

Marx also adopted Babbage's theory methodologically, including in *Capital*, where the chapter on machinery follows the chapter on the division of labour. There exists a structural homology between the design of machinery, and the division of labour, as Marx's argument highlights: 'The machine is a mechanism that, after being set in motion, performs with its tools the same operations as the worker formerly did with similar tools.'[32] In a footnote, he refers to Babbage's synthetic definition of machine ('The union of all these simple instruments, set in motion by a motor, constitutes a machine') and offers his own paraphrase:

The machine, which is the starting-point of the industrial revolution, replaces the worker, who handles a single tool, by a mechanism operating with a number of similar tools and set in motion by a single motive power, whatever the form of that power.[33]

It is at this point of *Capital* that Marx advances a further analytical principle that would go on to have an enormous influence on the methodology of the history of science and technology in the

---

31 Marx, *Grundrisse*, 704.
32 Marx, *Capital*, 495.
33 Ibid., 497.

twentieth century.³⁴ After challenging the belief that science, rather than labour, is the origin of the machine, Marx reverses the perception of the steam engine as the prime catalyst of the Industrial Revolution. Instead, he contends that it is the growth of the division of labour, its tools and 'tooling machines', that 'requires a mightier moving power than that of man', a source of energy that will be found in steam.³⁵ It was not the invention of the steam engine (*means of production*) that triggered the Industrial Revolution (as it is popular to theorize in the ecological discourse), but rather the developments of capital and labour (*relations of production*) demanding a more powerful source of energy:³⁶

> The steam-engine itself, such as it was at its invention during the manufacturing period at the close of the seventeenth century, and such as it continued to be down to 1780, did not give rise to any industrial revolution. It was, on the contrary, the invention of [tooling] machines [*Werkzeugmaschinen*] that made a revolution in the form of steam-engines necessary.³⁷

The 'mechanical monster' of the industrial factory was summoned first by labour and then accelerated by steam power, not the other way around.³⁸ Marx was clear: the genesis of technology is an *emergent process* driven by the division of labour. It is from the materiality of collective labour, from conscious and unconscious forms of

---

34 See Gideon Freudenthal and Peter McLaughlin (eds), *The Social and Economic Roots of the Scientific Revolution: Texts by Boris Hessen and Henryk Grossmann*, Berlin: Springer, 2009.

35 'An increase in the size of the machine and the number of its working tools calls for a more massive mechanism to drive it; and this mechanism, in order to overcome its own inertia, requires a mightier moving power than that of man.' Marx, *Capital*, 497.

36 Marx is mistakenly considered a *techno-determinist* for the prominence he grants to machinery in capitalism, but if he is determinist at all, he is a determinist of the *relations of production* and not of the *means of production*, as the division of labour, and not technology, is the driving force of capital. 'The inclusion of labor power as a force of production thus admits conscious human agency as a determinant of history: it is people, as much as or more than the machine, that make history.' Donald MacKenzie, 'Marx and Machine', *Knowing Machines*, Cambridge, MA: MIT Press, 1998, 26.

37 Marx, *Capital*, 496.

38 Ibid., 507.

cooperation, that extended apparatuses of machines emerge. Here, intelligence resides in the ramifications of human cooperation rather than in individual mental labour. Machine intelligence mirrors, embodies, and amplifies the analytical intelligence of collective labour.[39]

The alienation of knowledge by machinery

'What distinguishes the worst architect from the best of bees is that the architect builds the cell in his mind before he constructs it in wax.'[40] This is Marx's recognition, in *Capital*, of labour as a mental and individual activity. The collective division of labour, or labour in common, however, remains the *political inventor* of the machine.[41] A process of alienation of skill and knowledge starts as soon as machinery appears in front and in place of labour. Tools pass from the hands of the worker to the hands of the machine, and the same process happens to workers' knowledge: 'Along with the tool, the skill of the worker in handling it passes over to the machine.'[42] As such, the machine is but a crystallisation of collective knowledge. Marx condemns this alienation of the human mind, seconding Owen:

> Since the general introduction of soulless mechanisms in British manufactures, people have with rare exceptions been treated as a secondary and subordinate machine, and far more attention has been given to the perfection of the raw materials of wood and metals than to those of body and spirit.[43]

The introduction of machinery marks a dramatic dialectical turn in the history of labour, whereby the worker ceases to be the *subject* of the machine and becomes the *object* of capital: 'The hand tool makes the worker independent – posits him as proprietor. Machinery – as fixed

---

[39] For the idea of analytical intelligence, see Lorraine Daston, 'Calculation and the Division of Labour, 1750–1950', *Bulletin of the German Historical Institute* 62 (Spring 2018): 9–30.

[40] Marx, *Capital*, 284.

[41] Hodgskin gave great importance to observation (i.e., mental design) in the invention of machinery.

[42] Marx, *Capital*, 545.

[43] Robert Owen, 'Essays on the Formation of the Human Character' (1840), as quoted in Marx, *Grundrisse*, 711.

capital – posits him as dependent, posits him as appropriated.'[44] This shift in power between human and machine in the Victorian age is also the inception of a new imagery, in which machines acquire features of the living and the workers those of automata:[45]

> [It] is the machine which possesses skill and strength in place of the worker, is itself the virtuoso, with a soul of its own in the mechanical laws acting through it ... The worker's activity, reduced to a mere abstraction of activity, is determined and regulated on all sides by the movement of the machinery, and not the opposite. The science which compels the inanimate limbs of the machinery, by their construction, to act purposefully, as an automaton, does not exist in the worker's consciousness, but rather acts upon him through the machine as an alien power, as the power of the machine itself.[46]

The reflection on the alienation of knowledge from workers continues in *Capital*, where Marx has the process of knowledge extraction culminate in the full separation of science as a productive agent from labour:

> The knowledge, judgement and will which, even though to a small extent, are exercised by the independent peasant or handicraftsman, in the same way as the savage makes the whole art of war consist in the exercise of his personal cunning, are faculties now required only for the workshop as a whole. The possibility of an intelligent direction of production expands in one direction, because it vanishes in many others. What is lost by the specialized workers is concentrated in the capital which confronts them. It is a result of the division of labour in manufacture that the worker is brought face to face with the intellectual potentialities [*geistige Potenzen*] of the material process of production as the property of another and as a power which rules

---

44 Marx, *Grundrisse*, 702. Marx also quotes Hodgskin at 709: 'As soon as the division of labour is developed, almost every piece of work done by a single individual is a part of a whole, having no value or utility of itself. There is nothing on which the labourer can seize: this is my produce, this I will keep to myself.'

45 Simon Schaffer, 'Babbage's Dancer and the Impresarios of Mechanism', in *Cultural Babbage*, ed. Francis Spufford and Jenny Uglow, London: Faber & Faber, 1997, 53–80.

46 Marx, *Grundrisse*, 692–3.

over him. This process of separation starts in simple co-operation, where the capitalist represents to the individual workers the unity and the will of the whole body of social labour. It is developed in manufacture, which mutilates the worker, turning him into a fragment of himself. It is completed in large-scale industry, which makes science a potentiality for production which is distinct from labour and presses it into the service of capital.[47]

Marx comments upon the latter passage from *Capital* with a footnote to Thompson's *Inquiry into the Principles of the Distribution of Wealth* which is necessary to repeat:

'The man of knowledge and the productive labourer come to be widely divided from each other, and knowledge, instead of remaining the handmaid of labour in the hand of the labourer to increase his productive powers ... has almost everywhere arrayed itself against labour.' 'Knowledge' becomes 'an instrument, capable of being detached from labour and opposed to it.'[48]

Thompson provides a definition of knowledge labour that predates the twentieth-century theorists of the knowledge society and cognitive labour. As seen in the previous chapter, Thompson always included in the definition of labour 'the quantity of knowledge requisite for its direction' without which labour 'would be no more than brute force'.[49] In a polemic typical of Owenism, Thompson described machinery humiliating the 'general intellectual powers' of the workers, who were reduced to 'drilled automata'. Accordingly, the factory is an apparatus to keep the workers 'ignorant of the secret springs which regulated the machine and to repress the general powers of their minds' so 'that the fruits of their own labors were by a hundred contrivances taken away from them'.[50] In different passages, Thompson used the expressions 'general intellect', 'general intellectual power', 'general knowledge', and 'general power of the minds' in direct resonance with identical or equivalent terms used

---

47 Marx, *Capital*, 482.
48 Ibid., 483. Thompson, *Principles of the Distribution of Wealth*, 274.
49 Thompson, *Principles of the Distribution of Wealth*, 272.
50 Ibid., 292.

by Marx in the *Grundrisse*, such as 'general social labour', 'general scientific labour', 'general productive forces of the human brain', 'general social knowledge', and 'social intellect'.[51] Importantly, as indicated in the opening epigraph to this chapter, Thompson drew a direct link between the construction of a primarily white male general intellect and issues of gendered and racial discrimination. In Thompson's view, people are racist and chauvinist due to the lack of proper knowledge and education:

> Why also, it may be asked in reply, has the slavery of the blacks, and of women, been established? Because the whites in the one case, because the men in the other, made the laws: because knowledge had not been obtained on these subjects, the whites and the men erroneously conceiving it to be their interest to oppress blacks and women.[52]

Marx, for his part, also recognised the psychopathologies of industrial labour and the tactics to keep the workforce as illiterate as possible. He quoted Adam Smith's mentor, the Scottish philosopher Adam Ferguson, who had reached this conclusion a century earlier:

> Ignorance is the mother of industry as well as of superstition. Reflection and fancy are subject to err; but a habit of moving the hand or the foot is independent of either. Manufactures, accordingly, prosper most where the mind is least consulted, and where the workshop may . . . be considered as an engine, the parts of which are men.[53]

This should serve to remind us that the public mythology of artificial intelligence has always operated on the side of capital together with a hidden agenda to foster human stupidity, including the promulgation of racist and sexist ideologies.

---

51 Ibid., 272–362.
52 Ibid., 303.
53 Adam Ferguson, as quoted by Marx in *Capital*, 483. Marx cites Ferguson also for recognising as early as 1767 that 'thinking itself, in this age of separations, may become a peculiar craft'. Marx, *Capital*, 484.

The devaluation of capital by knowledge accumulation

What is the economic value of knowledge and science? What role do they play in capitalist accumulation? Marx explored these questions in an age that was flourishing with mechanical ingenuity, technical intelligence, and large infrastructures, such as railway and telegraph networks. In the passage on the general intellect, Marx considered knowledge in three ways: first, as a 'direct force of production' (*unmittelbaren Produktivkraft*); second, under the form of the 'social forces of production' (*gesellschaftlichen Produktivkräfte*); and third, as social practice (*gesellschaftlichen Praxis*), which is not abstract knowledge per se:

> Nature builds no machines, no locomotives, railways, *electric telegraphs*, *self-acting mules* etc. These are products of human industry; natural material transformed into organs of the human will over nature, or of human participation in nature. They are *organs of the human brain, created by the human hand*; the power of knowledge, objectified. The development of fixed capital indicates to what degree general social *knowledge* has become a *direct force of production*, and to what degree, hence, the conditions of the process of social life itself have come under the control of the *general intellect* and been transformed in accordance with it. To what degree the powers of social production have been produced, not only in the form of knowledge, but also as immediate organs of social practice, of the real life process.[54]

The general intellect becomes a transformative agent of society in a way that clearly echoes Thompson's optimism about the 'distribution of knowledge' as conducive to 'voluntary equality in the distribution of wealth'. The 'Fragment on Machines' contains an unresolved tension between *knowledge objectified in machinery* (as 'development of fixed capital') and *knowledge expressed by social production* (as 'development of the social individual'). Marx considers the primacy of knowledge in the production process and, then, the primacy of praxis over knowledge itself. The same thesis emerges in *Capital*, where Marx registers the stress of industrial labour on the workers' nervous system. Marx

---

54 Marx, *Grundrisse*, 706. The terms marked by asterisks appear in English in the original manuscript.

compares the economic value of individual skill against that of science. A realistic competition between the two is unlikely, since after a long process of 'separation of the intellectual faculties', the special skills of the worker vanish before the magnitude of the science, natural energy, and social labour that animates machinery:

> The separation of the intellectual faculties of the production process from manual labour, and the transformation of those faculties into powers exercised by capital over labour, is . . . finally completed by large-scale industry erected on the foundation of machinery. The special skill of each individual machine-operator, who has now been deprived of all significance, vanishes as an infinitesimal quantity in the face of the science, the gigantic natural forces, and the mass of social labour embodied in the system of machinery, which, together with those three forces, constitutes the power of the 'master'.[55]

In the 'Fragment', we have not only the recognition of knowledge as an alien power embodied in machinery (as found in Thompson) but also the attempt to assess the magnitude of its valorisation (which is missing in the latter). Here, Marx uses a criterion to assess knowledge accumulation that derives from the work of Thomas Hodgskin – a Ricardian socialist introduced in the previous chapter – often quoting his book *Popular Political Economy* (1827) and also praising his *Labour Defended against the Claims of Capital* (1825). Hodgskin pitted a positive emphasis on fixed capital as a concrete accumulation of past labour, knowledge, and science against the 'fiction' of circulating capital. In the *Grundrisse*, there is an echo of Hodgskin's ideas in Marx's claim that machinery is the 'most adequate form of fixed capital':

> The accumulation of knowledge and of skill, of the general productive forces of the social brain, is thus absorbed into capital, as opposed to labour, and hence appears as an attribute of capital, and more specifically of fixed capital, in so far as it enters into the production process as a means of production proper. Machinery appears, then, as the

---

55 Marx, *Capital*, 549.

most adequate form of fixed capital, and fixed capital . . . appears as the most adequate form of capital as such.⁵⁶

Modernising the Baconian motto 'Knowledge is power', authors of the industrial age such as Babbage, Thompson, and Hodgskin argued that knowledge is, without doubt, a productive and economic force. For Hodgskin, as much as for Thompson, it should be repeated, labour is primarily mental labour – that is, knowledge. 'Mental labour' is

> the labour of observing and ascertaining by what means the material world will give us the most wealth . . . Unless there be mental labour, there can be no manual dexterity; and no capability of inventing machines. It therefore is essential to production.⁵⁷

Importantly, for Hodgskin, there are neither intellectual hierarchies, nor division of hand and mind, nor a labour aristocracy in need of promotion: 'both mental and bodily labour are practised by almost every individual.'⁵⁸ In fact, Marx quotes Hodgskin in *Capital* to stress that skill is a common resource which is shared among workers and passes from one generation to the next.⁵⁹ Here, knowledge is a power that is collectively produced and shared, and this power constitutes (together with machinery and infrastructures) the core of fixed capital that must be reappropriated by workers (against the 'fiction' of circulating capital).⁶⁰

The most visionary passages of the *Grundrisse* refer to the crisis of capitalism due to the crisis of the centrality of labour, and therefore of the labour theory of value – which is to say, due to the fact that 'direct labour and its quantity disappear as the determinant principle of production . . . compared to general scientific labour, technological application of natural sciences . . . and to the general productive force arising from social combination [*Gliederung*]'.⁶¹ Further, says Marx:

---

56 Marx, *Grundrisse*, 694.
57 Hodgskin, *Popular Political Economy*, 45, 47.
58 Ibid., 47.
59 'Easy labour is transmitted skill.' Ibid., 48.
60 'Hodgskin called circulating capital a "fiction". Fixed capital was the stored-up skill of past labour.' Maxine Berg, *The Machinery Question and the Making of Political Economy*, Cambridge: Cambridge University Press, 1980, 274.
61 Marx, *Grundrisse*, 700.

Capital itself is the moving contradiction, [in] that it presses to reduce labour time to a minimum, while it posits labour time, on the other side, as sole measure and source of wealth ... On the one side, then, it calls to life all the powers of science and of nature, as of social combination and of social intercourse, in order to make the creation of wealth independent (relatively) of the labour time employed on it. On the other side, it wants to use labour time as the measuring rod for the giant social forces thereby created, and to confine them within the limits required to maintain the already created value as value. Forces of production and social relations – two different sides of the development of the social individual – appear to capital as mere means, and are merely means for it to produce on its limited foundation. In fact, however, they are the material conditions to blow this foundation sky-high.[62]

What looks like a contradiction in Marx's system (the obliteration of the political centrality of labour) is in fact the consequence of such centrality. Everywhere in the world, workers have been working enough! They have been producing so much and for so long that their past accumulated labour (under the forms of machinery, infrastructures, and collective knowledge) affects the rate of profit and slows down the economy. This is the thesis of the productivity of labour pitted against the unproductivity of capital, found in Hodgskin's *Labour Defended against Capital*. Marx, for his part, tries to prove that the accumulation of fixed capital (as machinery, infrastructures, collective knowledge, and science) could have profound side effects on the side of circulating capital (beside the chance of an overproduction crisis). In the *Grundrisse*, he accordingly explores the hypothesis that a growth of collective and technical knowledge could undermine capital's dominance, as Thompson and Hodgskin envisioned. Ultimately, in *Capital*, the utopian enthusiasms of the *Grundrisse* are reabsorbed by a realistic calculation of relative surplus value, which is adopted as the metrics of machinery and implicit metrics of knowledge value as well.

---

62 Ibid., 706.

The rise of the collective worker

In *Capital*, Marx replies to the Machinery Question by casting an extended social actor, the collective worker (*Gesamtarbeiter*), at the centre of the industrial theatre, whereas, for the bourgeoisie, it was an engineer with a steam engine. The figure of the collective worker replaces the personality cult of the inventor (individual mental labour) but also the idea of the general intellect (collective mental labour). Drawing on Babbage's labour theory of the machine, which explains the machine as the embodiment of the division of labour, Marx asserts the collective worker as the true *political inventor* of technology. The ambiguous hypothesis of the *knowledge theory of value* of the *Grundrisse* is thus finally grounded on an empirical basis: intelligence is logically materialised in the ramifications of the division of labour. The collective worker is a personification of the general intellect and, precisely, of its mechanisation.

Marx follows closely Babbage's labour theory of the machine in both the *Grundrisse* and *Capital*, but only in the latter does he make use of Babbage's principle of surplus labour modulation, which helps him to sketch the concept of relative surplus value and to measure the productivity of labour and machinery. Babbage's principle as quoted by Marx is as follows:

> The master manufacturer, by dividing the work to be executed into different processes, each requiring different degrees of skill or of force, can purchase exactly that precise quantity of both which is necessary for each process; whereas, if the whole work were executed by one workman, that person must possess sufficient skill to perform the most difficult, and sufficient strength to execute the most laborious of the operations into which the art is divided.[63]

Marx reverses the mystification of 'the master manufacturer' by restoring to the centre of the Babbage principle the collective worker who, needless to say, becomes now the main actor of the division of labour. The collective worker acquires features of a super-organism:

---

63 Marx, *Capital*, 469. The Penguin edition wrongly says 'Ch. 19, pp. 175' of Babbage's book: it is chapter 18, page 137.

The collective worker, formed out of the combination of a number of individual specialized workers, is the item of machinery specifically characteristic of the manufacturing period ... In one operation he must exert more strength, in another more skill, in another more attention; and the same individual does not possess all these qualities in an equal degree ... After the various operations have been separated, made independent and isolated, the workers are divided, classified and grouped according to their predominant qualities ... The collective worker now possesses all the qualities necessary for production in an equal degree of excellence, and expends them in the most economical way by exclusively employing all his organs, individualized in particular workers or groups of workers, in performing their special functions.[64]

In Marx's language, the collective worker becomes an 'item of machinery', a 'social mechanism', a 'collective working organism'.[65] Vivid machinic metaphors accompany the reincarnation of the general intellect as collective worker. The prehistory of the cyborg can be read between the lines of *Capital*:

The social mechanism of production, which is made up of numerous individual specialized workers, belongs to the capitalist ... Not only is the specialized work distributed among the different individuals, but the individual himself is divided up, and transformed into the automatic motor of a detail operation.[66]

The 'Fragment on Machines' emphasised not only the growing economic role of knowledge and science but also the role of social cooperation – that is, the growing role of the *general machinery* of social relations beyond the factory system. In a movement that resembles that of the construction of the *Gesamtarbeiter* within the factory, in the *Grundrisse* Marx sets 'the social individual ... as the great foundation-stone of production and of wealth' in the society to come:

---

64 Marx, *Capital*, 468–9.
65 See Henning Schmidgen, '1818: Der Frankenstein-Komplex', afterword to Bruno Latour, *Aramis: oder Die Liebe zur Technik*, trans. Gustav Roßler, Heidelberg: Mohr Siebeck, 2018, 303–19.
66 Marx, *Capital*, 481.

[The worker] steps to the side of the production process instead of being its chief actor. In this transformation, it is neither the direct human labour he himself performs, nor the time during which he works, but rather the appropriation of his own general productive power, his understanding of nature and his mastery over it by virtue of his presence as a social body – it is, in a word, the development of the social individual which appears as the great foundation-stone of production and of wealth.[67]

It seems that, with the transmutation of the general intellect into the collective worker, Marx also abandons the theory of capitalism's implosion due to the overproduction of knowledge as fixed capital. Capitalism will no longer collapse due to the accumulation of knowledge, because knowledge itself helps new apparatuses to improve the extraction of surplus value. Marxist scholar Michael Heinrich has noted that in *Capital*, 'when dealing with the production of relative surplus value, we can find an implicit critique of the "Fragment on machines" '.[68] Here, Marx appears to employ Babbage's principle of the modulation of surplus labour to design a theory of relative surplus value that recognises capitalism's capacity to maintain exploitation in equilibrium. According to Marx, surplus value can be augmented not just by reducing wages and material costs but also by increasing the productivity of labour in general – that is, by redesigning the division of labour and machines. If, according to Babbage's principle, the division of labour is an apparatus to modulate regimes of skill and therefore different regimes of salary according to skill, the division of labour becomes a modulation of relative surplus value. Being itself an embodiment of the division of labour, the machine then becomes the apparatus to discipline labour and regulate the extraction of relative surplus value.[69] As in Babbage's vision, the machine becomes a calculating engine – in this case, an instrument for the measurement of surplus value.

---

67 Marx, *Grundrisse*, 705.
68 Heinrich, 'Fragment on Machines', 197.
69 'One great advantage which we may derive from machinery is from the check which it affords against the inattention, the idleness, or the dishonesty of human agents.' Babbage, *On the Economy of Machinery*, 54.

The machine is a social relation, not a thing

In the twentieth century, Harry Braverman was probably the first Marxist to rediscover Babbage's pioneering experiments in computation and influence on Marx's theory of the division of labour.[70] While Marx read Thompson, Hodgskin, and Babbage, he never employed the notion of mental labour, probably in order to avoid supporting a labour aristocracy of skilled artisans as a political subject separate from the working class. For Marx, labour is always collective: there is no individual labour that is more prestigious than others, and, therefore, mental labour is always general; the mind is by definition social. Rather than a *knowledge theory of labour* that grants primacy to conscious activity, like the one in Thompson and Hodgskin, Marx maintains a *labour theory of knowledge* that recognises the cognitive import of forms of labour that are social, distributed, spontaneous, and unconscious. Intelligence emerges from the abstract assemblage of workers' simple gestures and micro-decisions, even and especially those which are unconscious.[71] In the general intellect studies and the history of technology, these are the in-between worlds of collective intelligence and unconscious cooperation, but also those of 'mechanised knowledge' and 'mindful mechanics'.[72] It ends up being Babbage who provides Marx with an operative paradigm to overcome Hegel's *Geist* and imbricate knowledge, science, and the general intellect into production.

As already stressed, the distinction between manual and mental labour disappears in Marxism because, from the abstract point of view of capital, all waged labour, without distinction, produces surplus value; all labour is abstract labour. However, the abstract eye of capital that regulates the labour theory of value employs a specific instrument to measure labour: the clock. In this way, what looks like a universal law

---

70 Harry Braverman, *Labor and Monopoly Capital: The Degradation of Work in the Twentieth Century*, New York: Monthly Review Press, 1974.

71 For the notion of the micro-decision, see Romano Alquati, 'Composizione organica del capitale e forza-lavoro alla Olivetti', part 2, *Quaderni Rossi* 3 (1963). Partially translated in Matteo Pasquinelli, 'Italian Operaismo and the Information Machine', *Theory, Culture, and Society* 32, no. 3 (2015): 55.

72 What in the following century will become the core of operationalism: management, logistics, and computer science. See Sandro Mezzadra and Brett Neilson, *The Politics of Operations: Excavating Contemporary Capitalism*, Durham, NC: Duke University Press, 2019.

has to deal with the metrics of a very mundane technology; after all, clocks are not universal.[73] Machines can impose a metrics of labour other than time, as has recently happened with social data analytics. As much as new instruments define new domains of science, likewise they define new domains of labour after being invented by labour itself.[74] Any new machine is a new configuration of space, time, and social relations, and it projects new metrics of such diagrams.[75] In the Victorian age, a metrology of mental labour existed only in an embryonic state. A rudimentary econometrics of knowledge begins to emerge only in the twentieth century with the first theory of information. The thesis of this chapter is that Marx's labour theory of value did not resolve the metrics for the domains of knowledge and intelligence, which had to be explored in the articulation of the machine design and in the Babbage principle.

Following Braverman and Schaffer, one could add that Babbage provided not just a labour theory of the machine but a *labour theory of machine intelligence*.[76] Indeed, Babbage's calculating engines ('intelligent machines' of their age) were an implementation of the analytical eye of the factory's master. Cousins of Jeremy Bentham's panopticon, they were instruments, simultaneously, of surveillance and measurement of labour. It is this idea that we should consider and apply to the age of artificial intelligence and its political critique, although reversing its polarisation, in order to declare computing infrastructures a concretion of labour in common.[77]

---

73 See Antonio Negri, *Time for Revolution*, London: Continuum, 2003, 27.

74 See Peter Damerow and Wolfgang Lefèvre, 'Tools of Science', in *Abstraction and Representation: Essays on the Cultural Evolution of Thinking*, ed. Peter Damerow, Dordrecht: Kluver, 1996, 395–404.

75 The idea that each machine establishes its own labour unit of measure constitutes a *machine theory of labour*, which cannot be expanded on here.

76 See Schaffer, 'Babbage's Intelligence'.

77 See Antonio Negri, 'The Re-Appropriation of Fixed Capital: A Metaphor?', in *Digital Objects, Digital Subjects*, ed. David Chandler and Christian Fuchs, London: University of Westminster Press, 2019, 205–14; Frederic Jameson, *An American Utopia: Dual Power and the Universal Army*, London: Verso, 2016.

5
The Abstraction of Labour

Cybernetics recomposes globally and organically the functions of the general worker that are pulverized into individual micro-decisions: the 'bit' links up the atomized worker to the 'figures' of the 'Plan'.

Romano Alquati, 1963[1]

The bifurcation of energy and information

With unusual insight, the French philosopher Gilbert Simondon once challenged the common understanding of the industrial age. He wrote: 'The industrial modality appears when the source of information and the source of energy separate, namely when the Human Being is merely the source of information, and Nature is required to furnish the energy. The machine is different from the tool in that it is a relay: it has two different entry points, that of energy and that of information.'[2] After all, the appearance of a modern notion of information is usually associated

---

1 'La cibernetica ricompone globalmente e organicamente le funzioni dell'operaio complessivo polverizzate nelle microdecisioni individuali: il "bit" salda l'atomo operaio alle "cifre" del "Piano".' Romano Alquati, 'Composizione organica del capitale e forza-lavoro alla Olivetti', part 2, *Quaderni Rossi* 3 (1963): 134, my translation.

2 Gilbert Simondon, 'Technical Mentality', *Parrhesia* 7 (2009): 20, originally published as 'Mentalité technique', *Revue Philosophique de la France et de l'étranger* 131, no. 3 (2006): 343–57.

with mass media such as press, telegraph, radio, and television – surely not with an industrial machine. Instead, Simondon proposed to see the industrial machine already as an *info-mechanical relay* because, for the first time, it was separating labour into energy (which was provided by natural resources such as water and coal) and information (the conscious movements and instructions of workers supervising and controlling the machine). According to Simondon's understanding, the traditional tool would be a design in which energy and information are still united: with the hammer, for example, the preindustrial artisan was still giving form and motion in the same gesture. This premodern unity of hand and mind was to be systematically disrupted by the industrial division of labour, as has been seen in chapter 3 revisiting the Machinery Question. Although it is often repeated that automation replaces labour, Simondon illustrated how automation actually displaces labour and bifurcates it into opposing lineages and hierarchies of manual and mental skill.

Writing at the end of the 1950s, in the France of the economic boom, Simondon was reading the industrial age under the influence of new technologies such as cybernetics which were also giving momentum to post-war modernist expectations.[3] Information was then an emerging notion that also affected the definition of labour. More recently, the Swedish scholar Andreas Malm has framed the industrial age from a different angle, given the current concerns about climate change and environmental degradation: the economy of energy.

Malm has argued, understandably, that the rise of the industrial mode of production was propelled by a stable and versatile form of energy, which was found in coal after the use of waterpower.[4] But according to him, coal contributed to the acceleration of industrial capitalism not just because of its energetic potential, but because its physical properties, such as lightness, homogeneity, and measurability, matched perfectly the new abstract dimensions of capital. Steam engines replaced water mills not because coal was cheaper and more abundant than water, but because it provided a more stable flow of power than rainfalls and allowed factories to move close to urban areas, where the working class

---

3 See Henning Schmidgen, 'Thinking Technological and Biological Beings: Gilbert Simondon's Philosophy of Machines', *Revista do Departamento de Psicologia UFF* 17 (2005): 11–18.

4 Andreas Malm, 'The Origins of Fossil Capital: From Water to Steam in the British Cotton Industry', *Historical Materialism* 21, no. 1, (2013): 15–68.

was living at the time. Malm has registered in this way the energetic rationale for the slow emergence of fossil capitalism out of the manufacturing age. It took roughly forty years for the steam engine to be adopted in the place of the water mill: if coal came to be used across the full spectrum of production, it was because it was the most adequate source of *abstract energy*, where 'abstract' means easily computable in terms of cost, transport, stock, performance, and social organisation.

Moreover, it was also thanks to the steam engine that coal was turned into a key component of industrial capitalism, but precisely because this technological innovation could turn its energy potential into a stable and continuous motion, as Malm has noticed:

> For coal to be universalised as a fuel for all sorts of commodity production, it had to be turned into a source of mechanical energy – and, more precisely, of rotary motion. Only by coupling the combustion of coal to the rotation of a wheel could fossil fuels be made to fire the general process of growth: increased production – and transportation – of all kinds of commodities. This is why James Watt's steam engine is widely identified as the fatal breakthrough into a warmer world.[5]

The rise of the steam engine and the adoption of coal, however, were not autonomous drivers of industrial development in their own: they were responding to more profound economic dynamics such as a new regime of labour exploitation. Capitalism brought about a need for a more streamlined – *more abstract* – system of organizing the workforce. This involved utilising the clock to accurately measure labour time and implementing a precise division of the labour space, all of which were made possible under the supervision of the factory's master. The energetic versatility of coal and the mechanical exactitude of steam engines helped consolidate the new spatio-temporal abstractions of industrial capitalism.

Extending Malm's analysis to include Simondon's insight, one may add that the abstract properties of information emerged together with these spatio-temporal abstractions and the new characteristics of fossil energy, i.e., its homogeneous carbon chains, which made coal easier to

---

5 Ibid., 18.

be quantified and computed than traditional sources, such as water and animal power. What should be recognised in the gears of the industrial machine, then, is the bifurcation and coupling at the same time of *abstract energy* (or standardised motion), and *abstract form* (or information), both understood as quantifiable operations and means of production.[6]

If human labour was separated into abstract energy and abstract forms, this was also thanks to two new technologies of control: feedback devices like James Watt's steam governor (1788) and controllers like the Jacquard loom's punched card (1801).[7] The steam governor was a device to maintain the constant output of an engine by regulating its fuel input in real time. The punched card was a data device to store instructions of textile patterns. To be more precise, Watt's governor turned engine impulses into *abstract motion* – that is, constant rotary motion – and Jacquard's punched cards turned manual instructions into *information* – that is, computable knowledge. These two devices can be considered retrospectively as the first cybernetic devices. Watt's governor was the first example of an exact information feedback system, while the Jacquard loom's punched card would be adopted by IBM as standard storage format, almost unchanged throughout the twentieth century.

A note on the controversy of abstract labour

It was Hegel who saw labour as an abstraction that forms machines and subjectivity and defined 'abstract labour' for the first time during his 1805–6 Jena lectures, interpreting most likely Adam Smith's passages on the division of labour and the invention of machines:

> Man's labor itself becomes entirely mechanical, belonging to a many-sided determinacy. But the more abstract [his labor] becomes, the more he himself is mere abstract activity. And consequently he is in a position to withdraw himself from labor and to substitute for his own

---

6 Abstract movement could be defined also as abstract labour, as Hegel did originally in the Jena lectures: see below. See also the critique of Sohn-Rethel in chapter 8.

7 James Beniger, *The Control Revolution: Technological and Economic Origins of the Information Society*, Cambridge, MA: Harvard University Press, 1986, 17.

activity that of external nature. He needs mere motion, and this he finds in external nature. In other words, pure motion is precisely the relation of the abstract forms of space and time – the abstract external activity, the *machine*.[8]

Later, in the *Philosophy of Right* (1820), Hegel kept defining abstract labour in a similar way: 'the abstraction of production makes work increasingly *mechanical*, so that the human being is eventually able to step aside and let a *machine* take his place.'[9] Marx subsequently hijacked Hegel's interpretation and declared that the social abstraction of concern was not only the division of labour but also wage labour, that is, the rule of capital over labour. Capitalism effectively turned human labour into an abstraction, but this abstraction was the commodity form. Under industrial capitalism, labour was quantified in abstract time units, rendered a general equivalent throughout society, and traded as a commodity like any other, in fact as the very substance of all commodities.[10] Marx saw the *abstraction of labour* primarily as a function of capital, as a wage relation.

During the industrial era, however, the process of abstraction affected space, time, energy, labour, and knowledge in different ways, and this power of abstraction cannot be considered an expression of technocapitalism alone. In fact, everyone lays claim to abstraction. But to whom does the power of abstraction truly belong, at the end? Who or what possesses the political agency that shapes the social abstractions of history?

---

8 Georg Wilhelm Friedrich Hegel, *Hegel and the Human Spirit: A Translation of the Jena Lectures on the Philosophy of Spirit (1805-6)*, ed. and trans. Leo Rauch, Detroit: Wayne State University Press, 1983, 121.

9 G. W. F. Hegel, *Elements of the Philosophy of Right*, ed. Allen W. Wood, trans. H. B. Nisbet, Cambridge Texts in the History of Political Thought, Cambridge: Cambridge University Press, 1991, sections 198, 232–3. See also the third part of the *Encyclopaedia of the Philosophical Sciences*: G. W. F. Hegel, *Hegel's Philosophy of Mind*, ed. and trans. William Wallace, Oxford: Clarendon Press, 1894, §526, 123.

10 Marx, *Capital*, 137.

The labour of information

The historian James Beniger argues that between the late nineteenth and mid-twentieth centuries, information technologies emerged because of the economic boom of Western countries and the need to govern industrial production and distribution. In other words, it was the economic acceleration which prompted the transformation of analogue media into numerical information. The genesis of the paradigms of cybernetics and information theory responded to a 'crisis of control' of Western capitalism that had to manage a commodity surplus and new large infrastructures of distribution. Rather than *information revolution*, as it is often styled, Beniger termed this development as the oxymoronic *control revolution* (which is a fitting description also of the historical drive of cybernetics for political equilibrium).[11] In order to govern a growing economy, a more abstract definition of information (that is measurable, computable, and transmissible knowledge) had to be introduced.

This historical process can be framed, once again, not just from the point of view of commodity circulation but the organisation of labour. Pace Beniger, Marxist accounts of the post-industrial age have stressed the role of labour conflicts and social struggles, rather than economic surplus, in prompting technological development. They have also contested the political neutrality of the technical notion of information, as the Italian sociologist Romano Alquati did, for instance, in his inquiry into labour conditions at the Olivetti computer factory in the early 1960s.[12]

Olivetti was a pioneering company famous for producing typewriters, electronic calculators, and mainframe computers from the 1950s. In 1959, Olivetti launched, for instance, the Elea 9003, the first transistor-based commercial computer, whose futuristic graphical user interface was designed by Ettore Sottsass. It was at the Olivetti factory in Ivrea that Alquati applied for the first time the method of workers' inquiry (or, *conricerca*) to the organisation of

---

11 Beniger, *Control Revolution*, 6.
12 Alquati, 'Composizione organica del capitale'. See, in particular, Matteo Pasquinelli, 'Italian Operaismo and the Information Machine', *Theory, Culture and Society* 32, no. 3 (2015): 49–68.

labour in cybernetics. Workers' inquiry was a sort of participatory action research, albeit more militant, and was based in the active involvement of workers, also with the extensive use of individual and group interviews.

Alquati discovered, then, that the cybernetic apparatus of the Olivetti factory first was an extension of its internal bureaucracy, which monitored workers at the assembly line and the production process in general by the means of 'control information'. It was via the circuits of cybernetics that bureaucracy was finally able to descend into the bodies of the workers and watch their activities closely. Although Alquati viewed cybernetics as an extension of bureaucracy, he reversed the top–down perspective that is implicit in the idea of control information. In addition to 'control information', he coined the term 'valorising information' to describe the flow of information that is generated by the workers and that, running upstream, feeds the circuits of the factory, and gives form to the final products. In this view, information is continuously produced by workers, absorbed by machinery, and eventually condensed into commodities:

> Information is essential to labour-force, it is what the worker – by the means of constant capital – transmits to the means of production on the basis of evaluations, measurements, and elaborations in order to operate on the object of work all those modifications of its form that give it the requested use value.[13]

With Alquati, numerical information enters, probably for the first time, the definition of labour. Alquati noticed that the most important part of labour is made by the series of creative acts, measurements, and decisions that workers constantly have to perform in front of the machine and in the assembly line. He called information precisely all the innovative 'micro-decisions' that workers take along the production process, that give form to the product, but also regulate the machinic apparatus itself:

---

13 'L'informazione è l'essenziale della forza-lavoro, è ciò che l'operaio attraverso il capitale costante trasmette ai mezzi di produzione sulla base di valutazioni, misurazioni, elaborazioni per operare nell'oggetto di lavoro tutti quei mutamenti della sua forma che gli danno il valore d'uso richiesto.' Alquati, 'Composizione organica del capitale', 121, my translation.

The *productive labour* is defined by the quality of *information* elaborated and transmitted by the worker to the *means of production* via the mediation of *constant capital*, in a way that is tendentially *indirect*, but completely *socialised*.[14]

According to Alquati, it is specifically the numerical dimension of cybernetics that can encode workers' knowledge into digital bits and, consequently, transform digital bits into numbers for economic planning (as stated in the opening epigraph to this chapter).[15]

Alquati saw an extended structure merging bureaucracy, management, cybernetic machinery, and the division of labour: this was a new system taking the place of the old factory's master. Cybernetics unveiled the machinic nature of bureaucracy and, conversely, the bureaucratic role of machines – that is, how they both work as feedback apparatuses to control and capture workers' know-how. The findings of Alquati's research can be summarised as follows: (1) labour is the source of information of the industrial cybernetic apparatus, indeed the most valuable part of labour is information; (2) information operates the cybernetic apparatus, gradually improves its design and adds value to the final products; (3) the numeric dimension of cybernetics allows us to translate labour into knowledge, knowledge into information, information into numbers, and so, numbers into economic planning; (4) the cybernetic apparatus of the factory grows and improves thanks to the contribution of workers' socialised intelligence. For the first time in a distinct way, the cybernetic or automated factory made visible the transformation of labour into measurable knowledge – that is, information.

In the early 1960s, Alquati and Italian *operaismo* started to register the transformation of Fordism and its more and more 'abstract' division of labour across society. This was clearly prefigured also by political philosopher Mario Tronti's image of the *social factory*. In 1962, Tronti wrote that 'at the highest level of capitalist development . . . the whole of society becomes an articulation of production, the whole society lives in function of the factory and the factory extends its exclusive dominion

---

14 'Il *lavoro produttivo si* definisce nella qualità delle informazioni elaborate e trasmesse dall'operaio ai *mezzi di produzione*, con la mediazione del *capitale costante* in modo tendenzialmente *indiretto*, ma completamente *socializzato*.' Alquati, 'Composizione organica del capitale', 121, my translation.

15 Alquati, 'Composizione organica del capitale', 134, my translation.

over the whole society'.¹⁶ Information technologies were the material infrastructure that innervated the regime of industrial capitalism into society. Although Italian *operaismo* always had a secondary interest in science and technology, Alquati gave a key contribution on this matter. He maintained that any technological innovation, including cybernetics, always embodies the power relations and class antagonism of a given historical moment and that for this reason it should be the focus of study:

> Capital is always accumulated social labour, the machine is always incorporated social labour. Obviously. Every 'new machine', every innovation expresses the general level and quality of the power relations between classes at that moment.¹⁷

In the end, it is not difficult to see the rise of information technologies as part of the long evolution of the spatio-temporal abstractions that have been disciplining labour power in the past century. Information came to measure the intelligence, knowledge, and skills needed to master the production process and social relations at large.¹⁸ Coincidentally, this meaning is not far removed from the origin of the term 'information' that was introduced to replace 'intelligence' in the early days of information theory. In 1928, the US engineer Ralph Hartley of the Bell Telephone Labs proposed to revise the act of 'intelligence' or 'interpretation of a signal', which were at that time expressions commonly used in telegraphy, with a notion devoid of any reference to human faculties and, essentially, measurable.¹⁹ This originary role of *human intelligence* in communication technologies can be taken as further evidence of information theory's interest in the automation and

---

16 Mario Tronti, *Workers and Capital*, London: Verso, 2019, 26.

17 'Il capitale è sempre lavoro sociale accumulato, la macchina è sempre lavoro sociale incorporato. Ovvio. Ogni "nuova macchina", ogni innovazione esprime il livello generale e la qualità dei rapporti di forza fra le classi in quel momento.' Alquati, 'Composizione organica del capitale', 89, my translation.

18 On energy and information as metrics of labour, see Matteo Pasquinelli, 'Labour, Energy, and Information as Historical Configurations: Notes for a Political Metrology of the Anthropocene', *Journal of Interdisciplinary History of Ideas* 11, no. 22 (2023).

19 Ralph V. Hartley, 'Transmission of Information', *Bell System Technical Journal* 7 (1928). See also: Bernard D. Geoghegan, 'Information', in *Digital Keywords: A Vocabulary of Information Society and Culture*, ed. Benjamin Peters, Princeton: Princeton University Press, 2016, 173–83.

deskilling of mental labour, but also as a confirmation of a trajectory that significantly has unfolded, after a long technological cycle, into the project of *artificial intelligence*. Nowadays, the 'intelligence' that AI algorithms encode and measure extends to an increasingly wide social field, as this book has attempted to show. This type of intelligence belongs to both manual and mental labour, to explicit and tacit knowledge, but above all to the capacity of cooperation and self-organisation, which is quintessentially a political craft. Going beyond the horizon of electromechanical engineering, what information comes ultimately to measure and mediate is the antagonism between workers and capital – the 'signals' that are exchanged between these two noisy camps of the social order.

PART II
The Information Age

6
The Self-Organisation of the Cybernetic Mind

It has been rightly urged that a history of brain models is really a history of the literary and material technologies which are familiar to, and then used as metaphors by, brain scientists. Their metaphorical menagerie exhibits mental clocks, logical pianos, barrel organisms, neural telegraphs and cerebral computer nets. How do specific technologies get into this zoo? Claims that certain systems can mimic, or even exhibit, intelligence are sustained by social hierarchies of head and hand. Minds are known because these social conventions are known.

Simon Schaffer, 'OK Computer', 2001[1]

[The] wonder of our time, electrical telegraphy, was long ago modeled in the animal machine. But the similarity between the two apparatus, the nervous system and the electric telegraph, has a much deeper foundation. It is more than similarity; it is a kinship between the two, an agreement not merely of the effects, but also perhaps of the causes.

Emil Du Bois-Reymond, *On Animal Motion*, 1851[2]

---

1 Simon Schaffer, 'OK Computer', in *Ansichten der Wissenschaftsgeschichte*, ed. Michael Hagner, Frankfurt: Fischer, 2001, 393–429.

2 Emil Du Bois-Reymond, 'Über thierische Bewegung' (1887), 2 vols., translated by Laura Otis in 'The Metaphoric Circuit: Organic and Technological Communication in the Nineteenth Century', *Journal of the History of Ideas* 63, no. 1 (2002): 105.

> *The nervous systems ... have been externalized, as part of the reversal of the interior and exterior worlds. Highways, office blocks, faces and street signs are perceived as if they were elements in a malfunctioning central nervous system.*
>
> J. G. Ballard, The Atrocity Exhibition, 1990[3]

A social history of the nervous system

In 2012, the AlexNet algorithm – a large artificial neural network – won the ImageNet competition, which is the international benchmark for image recognition software. Since then, 'deep' artificial neural networks, also known as 'deep learning', have led the machine learning revolution and have been regarded as the most effective technique of AI. Their success revived also expectations that the 'solution' to AI may be found in the secret logic of the brain's structures – an idea that dates back to the early days of digital computers. Neurophysiologist Warren McCulloch and mathematician Walter Pitts were the first to propose imitating biological neurons in a device.[4] In their 1943 paper 'A Logical Calculus of the Ideas Immanent in Nervous Activity', they presented artificial neural networks as an imitation of the brain's physiology, but their idea concealed also an external 'social' genealogy that this chapter intends to rediscover and excavate.[5] Rather than as a biomorphic artefact (i.e., an artefact imitating life forms), this chapter proposes to illuminate artificial neural networks from a different and unusual perspective – that is, as a technique for the *self-organisation of*

---

3 J. G. Ballard, *The Atrocity Exhibition*, annotated ed., San Francisco: Re:Search, 1990, 44n.

4 Warren McCulloch and Walter Pitts, 'A Logical Calculus of the Ideas Immanent in Nervous Activity', *Bulletin of Mathematical Biophysics* 5 (1943): 115–33. Artificial neural networks were discussed as an architecture of computation well before the term 'artificial intelligence' was introduced. Turing began to speculate about intelligent machinery in 1947, while McCarthy coined the term 'artificial intelligence' in 1955 (see the introduction of this book). Alan Turing, 'Lecture to the London Mathematical Society', 20 February 1947, in *Collected Works*, vol. 1, North-Holland Publishing Company, 1985; Christof Teuscher, *Turing's Connectionism: An Investigation of Neural Network Architectures*, Berlin: Springer, 2012, viii.

5 Margaret Boden regards McCulloch and Pitts's paper as an 'abstract manifesto for computational psychology', a founding text for the both the lineages of connectionist and symbolic AI. Margaret Boden, *Mind as Machine: A History of Cognitive Science*, Oxford: Oxford University Press, 2006, 190.

information. This hypothesis aligns their invention with the labour theory of automation which was expounded in the first part of this book. As much as the design of industrial machines emerged from the imitation of the organisation of labour, similarly, artificial neural networks (and machine learning algorithms in general) can be considered as machines that self-organise their parameters – their internal design – imitating the organisation of the external world. Rather than an 'ontological theatre' of the living, as historian of science Andrew Pickering has defined them, cybernetic experiments of self-organisation were essentially a laboratory of the social.[6]

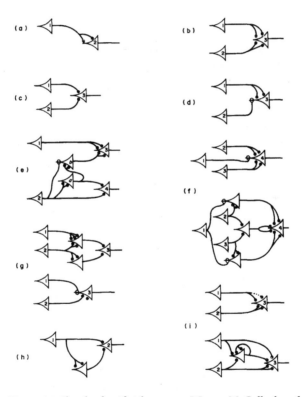

Figure 6.1. Sketch of artificial neurons. Warren McCulloch and Walter Pitts, 'A Logical Calculus of the Ideas Immanent in Nervous Activity', *Bulletin of Mathematical Biophysics* 5, no. 4 (1943): 105.

---

6 Andrew Pickering, *The Cybernetic Brain: Sketches of Another Future*, Chicago: University of Chicago Press, 2010; Andrew Pickering, 'A Gallery of Monsters: Cybernetics and Self-Organization, 1940–1970', in *Mechanical Bodies, Computational Minds*, ed. Stefano Franchi and Güven Güzeldere, Cambridge, MA: MIT Press, 2005, 229–45.

In 1943, interpreting laboratory findings in their own way, McCulloch and Pitts proposed to formalise the human brain as a 'nervous net' that performs logical operations (see fig. 6.1). They envisioned a network of computing nodes that could imitate human reasoning by reducing human logic to Boolean logic and its AND, OR, and NOT operators. The analogy between brain anatomy, logical inference, and computing devices was based on the observation that biological neurons display an 'all or none', or binary, behaviour. If the sum of the impulses which a neuron receives from its excitatory and inhibitory synapses exceeds a given limit, the neuron fires a signal to the synapsis of the following neuron; otherwise, it remains quiescent.[7] The novelty of the idea was not the *network form* per se but rather the *threshold logic* that, in such structure, impersonates the Boolean operators and the progressive steps of inferential reasoning. By adjusting the behaviour of their nodes, these machines were said to be 'learning' like brains – that is, to be recording complex information through their self-organisation.

A few years prior, in 1938, the US mathematician and cryptographer Claude Shannon had demonstrated that electric switching circuits could execute the Boolean logic operations. He designed the AND, OR, and NOT logic gates, which soon became incorporated in all transistors and microchips, thus laying the foundation of the computer age.[8] The emergence of neural networks as a key idea for AI is best understood by examining Shannon's logic gates rather than brain physiology. Essentially, McCulloch and Pitts argued that the brain's neural circuits perform the same operations as Shannon's electrical circuits. While machine learning textbooks reiterate that McCulloch and Pitts's idea of artificial neurons was inspired by the structures and behaviour of neurons in the brain, in fact the opposite is true: they saw, in the first instance, biological neurons as technological artefacts. McCulloch and Pitts implicitly envisioned brain physiology as homologous with the communication technology of the age, comprised of electromechanical relays, feedback

---

7 'To see how neurons might represent logical assertions, consider two examples. Suppose that both A and B are excitatory fibres, and that a pulse travelling along either suffices to cross the synaptic gap. This physical situation corresponds to the formal statement, "If either A or B is true, then C is true."' Steve Heims, *The Cybernetics Group*, Cambridge, MA: MIT Press, 1991, 42.

8 Claude Shannon, 'A Symbolic Analysis of Relay and Switching Circuits', *Transactions of the American Institute of Electrical Engineers* 57, no. 12 (1938): 713–23.

mechanisms, television scanners, and, notably, telegraph networks. At the 1948 Hixon symposium on cerebral mechanisms, discussed in more detail in the next chapter, McCulloch urged his colleagues to 'conceive neurons as telegraphic relays'.[9]

To a historian of science and technology, McCulloch and Pitts's artificial neural networks appear not as a completely original idea but as an elaboration upon an old one. Laura Otis, for one, has shown that the analogy between the nervous system and electric networks was already established in the nineteenth century and drawn upon by, among others, the telegraph inventor Samuel Morse and physicist Hermann von Helmholtz.[10] As evidence of this intellectual climate, in an 1851 lecture on the subject of animal motion cited at the beginning of this chapter, the Berlin physiologist Emil Du Bois-Reymond expounded on the similarity between the nervous system and electric telegraph networks with a visionary fervour closer to science fiction than science.

The imprint of the infrastructures of communication extended beyond the 'neural telegraph' analogy of the nineteenth century: it can be found also in the twentieth century's cybernetic projects of self-organisation that remained crucial in the evolution of artificial neural networks. Indeed, the idea of self-organising computation capable of adapting to the environment and 'learning' in enduring fashion is a key part of the 'epistemic ensemble' of cybernetics that paved the way to machine learning.[11] Owing to the academic hegemony of symbolic AI and the widespread anthropomorphisation of technology, it is hard to imagine contemporary AI as a technique of self-organising information or 'spontaneous order' emerging out of data. And yet, this is a realistic description of what machine learning actually does. The link between the self-organising computation of twentieth-century and twenty-first-century AI has been concealed by a complex stratum of technological advancements, in which we lost sight of its origin and development. This chapter undertakes a 'dig' into this

---

9 Lloyd A. Jeffress (ed.), *Cerebral Mechanisms in Behavior: The Hixon Symposium*, New York: Wiley, 1951, 45.

10 See Laura Otis, 'The Metaphoric Circuit: Organic and Technological Communication in the Nineteenth Century', *Journal of the History of Ideas* 63, no. 1 (2002): 105–28. See also Laura Otis, *Networking: Communicating with Bodies and Machines in the Nineteenth Century*, Ann Arbor: University of Michigan Press, 2001; and Christoph Hoffmann, 'Helmholtz' Apparatuses: Telegraphy as Working Model of Nerve Physiology', *Philosophia Scientiae* 7, no. 1 (2003): 129–49.

11 See also Michael Castelle, 'Deep Learning as an Epistemic Ensemble', 2018, castelle.org.

stratum – into a prehistory of machine learning in which social, communication, and computational networks were all part of continuous (and contiguous) movements of self-organisation.

Mechanising self-organisation

In the second half of the twentieth century, self-organisation rose as a popular topic across a wide range of disciplines, including biology, chaos theory, neuroscience, thermodynamics, and even neoliberal economics (if one considers the peculiar interest in the 'spontaneous order' of markets). How should we interpret such a widespread quest for the principles of self-organisation? The first impression is of a diverse movement searching for an ontological principle of life; however, such a quest for 'life' principles appears to mirror the 'principles of self-organisation' that could be detected also in the societal changes of the post-war period.

Originally, it was modern political philosophy (with Spinoza and Kant) which conceived of self-organisation and autonomy as key notions for theorising the social contract and individual freedom. But, for some reason, in the mid-twentieth century, the principle of self-organisation migrated from the social ontology and was transformed into an extra-social ideal for vitalist philosophies (with its highest manifestation in James Lovelock's Gaia hypothesis, which deems planet Earth a super-organism).[12] In 1977, Ilya Prigogine was awarded the Nobel Prize for his studies on self-organising structures in thermodynamic systems far from equilibrium.[13] In the same year, Langdon Winner's book *Autonomous Technology* signalled a further mutation in the discourse of self-organisation, whereby technology rather than nature was newly perceived to be 'autonomous' from the human and dangerously out of control, thus reviving certain Frankensteinian narratives of the industrial age.[14] As these examples show, the concept of self-organisation has accrued, across different centuries and disciplines, a thick ideological

---

12 J. E. Lovelock, 'Gaia as Seen through the Atmosphere', *Atmospheric Environment* 6, no. 8 (1972): 579–80.

13 Gregoire Nicolis and Ilya Prigogine, *Self-Organization in Non-Equilibrium Systems*, New York: Wiley, 1977.

14 Langdon Winner, *Autonomous Technology: Technics-Out-of-Control as a Theme in Political Thought*, Cambridge, MA: MIT Press, 1978.

patina. When and how exactly did the contemporary idea of self-organisation consolidate?

Curiously, as philosopher of biology Evelyn Fox Keller has noted, it took cybernetics, which is a branch of electromechanical engineering and not a natural science, to reboot the scientific debate on self-organisation in the twentieth century.[15] In the 1940s, cybernetics took over the modern dream of building 'thinking machines', albeit by adopting a different technique from the previous century. In the industrial era, Babbage had envisioned the automation of mental labour through an 'engine' that implemented hand calculation. Human reasoning was then encoded as a logical procedure, as a linear sequence of step-by-step operations (which Alan Turing would associate later with the telegraph tape to envision his eponymous machine). Cyberneticians explored other ways of building 'intelligent automata'. Rather than imitating the rules of human reasoning, they aimed at imitating the rules by which organisms organise themselves and adapt to the environment. Self-organisation was understood, importantly, also as self-reproduction and self-repair. This was a key aspect that Kant stressed in his definition of 'organic beings', which remained a guiding principle for the cyberneticians.[16]

Cybernetics claimed to have found in all organisms a basic 'mechanism' of behaviour – that is, information as a medium of feedback with the environment and internal self-regulation. In one of its founding texts, Arturo Rosenblueth, Norbert Wiener, and Julian Bigelow claimed that 'the broad classes of behaviour are the same in machines and in living organisms'.[17] Although different in narrow classes (organisms, obviously, do not have wheels, etc.), the article posited that both machines and organisms operate thanks to information feedback that shapes their purpose and teleology. This principle of cybernetics had, in fact, been anticipated in early-century biology by Jakob von Uexküll, who viewed

---

15 Evelyn Fox Keller, 'Organisms, Machines, and Thunderstorms: A History of Self-Organization', part 1, *Historical Studies in the Natural Sciences* 38, no. 1 (2008): 45–75.

16 'In such a natural product as this every part is thought as *owing* its presence to the agency of all the remaining parts, and also as existing *for the sake of the others* and of the whole, that is as an instrument, or organ . . . The part must be an organ *producing* the other parts – each, consequently, reciprocally producing the others . . . Only under these conditions and upon these terms can such a product be an *organized* and *self-organized being*, and, as such, be called a *physical end*.' Immanuel Kant, *Critique of Judgement*, 1790, quoted in Keller, 'Organisms, Machines, and Thunderstorms', part 1, 49.

17 Arturo Rosenblueth, Norbert Wiener, and Julian Bigelow, 'Behaviour, Purpose and Teleology', *Philosophy of Science* 10, no. 1 (1943): 18–24.

the organism as an information processing system struggling to adapt to its environment. Uexküll defined the exchange between an animal's nervous system (*Innenwelt*) and the outside world (*Außenwelt*, or *Umwelt*) as a 'function circle' (*Funktionskreis*). In Wiener's coinage, it should be remembered, the term 'cybernetics' (from the Greek *kybernetes*, or 'steersman') referred to the capacity of a technical, social, and living system to control itself via an exchange of information with the environment. It is quite evident that both Uexküll and cyberneticians derived from the communication systems of their age – the telegraph, telephone, and radio networks – an analogy for the interaction of living beings with their environment.

Although cyberneticians initially considered information under the form of analogue electromagnetic signals, they gradually shifted towards digital, discrete, and computational 'bits'.[18] For cybernetics, it was not simply machines in general but also digital computers (i.e., finite state automata) that could imitate the living being's principles of self-organisation.[19] The British psychiatrist Ross Ashby was the main theorist of self-organisation in cybernetics. His 1947 paper, 'Principles of the Self-Organising Dynamic System', was committed to demonstrating that self-organisation was not only a feature of the living but could be also of 'strictly determinate' machines, that is, computers:[20]

> It has been widely denied that a machine can be 'self-organizing', i.e., that it can be determinate and yet able to undergo spontaneous changes of internal organisation. The question of whether such can occur is not of purely philosophic interest for it is a fundamental problem in the theory of the nervous system. There is much evidence that this system is both (a) a strictly determinate physico-chemical system, and (b) that it can undergo 'self-induced' internal reorganisations resulting in changes of behaviour. It has sometimes been held that these two requirements are mutually exclusive. The purpose of this paper is to show that a machine

---

18 This is the subtle difference between the two information theories by Weiner and Shannon (analogue and digital, respectively). See Ronald Kline, *The Cybernetics Moment: Or Why We Call Our Age the Information Age*, Baltimore: Johns Hopkins University Press, 2015.

19 As Turing defined them, 'discrete state machines ... are the machines which move by sudden jumps or clicks from one quite definite state to another'. Alan Turing, 'Computing Machinery and Intelligence', *Mind* 59, no. 236 (October 1950), in *Essential Turing: Seminal Writings in Computing, Logic, Philosophy, Artificial Intelligence, and Artificial Life*, ed. B. Jack Copeland, Oxford: Oxford University Press, 446.

20 For a portrait of Ashby and his homeostat see Pickering, 'A Gallery of Monsters'.

can be at the same time (a) strictly determinate in its actions, and (b) yet demonstrate a self-induced change of organisation.[21]

Ashby put his theory into practice by inventing the 'homeostat' (whose name is a tribute to the homeostasis of living systems, as defined by Walter Bradford Cannon in 1926). Built from four bulky electro-mechanical units, however, its capacity of 'self-organisation' was rather the opposite of wilful adaptation to external stimuli. Grey Walter sarcastically called it *machina sopora* ('sleeping machine' in Latin), since 'its goal was to become quiescent; it changed state only when disturbed from outside'.[22] In a later article from 1960, 'Principles of the Self-Organizing System', Ashby fully ruled self-organisation under a mechanistic paradigm in a way to get rid of the metaphysics about life emergence: 'While, in the past, biologists have tended to think of organisation as something extra, something *added* to the elementary variables, the modern theory, based on the logic of communication, regards organisation as a restriction or constraint.' He concluded, however, in a way that is an illuminating example of the typical desire of automation to replace humanity and achieve invisibility: 'I think that in the future we shall hear the *word* ['organisation'] less frequently, though the *operations* to which it corresponds, in the world of computers and brain-like mechanisms, will become of increasingly daily importance.'[23]

Cyberneticians like Ashby did not pursue self-organisation simply as a key to the imitation of living structures but specifically of brain structures. Consequently, they studied the self-organisation of neural networks as the key to intelligent behaviour. A turning point in this debate came in 1949, when neuropsychologist Donald Hebb published a crucial book, *The Organisation of Behavior*, in which he claimed to have identified a basic rule of self-organisation in neural networks.[24]

---

21 Ross Ashby, 'Principles of the Self-organizing Dynamic System', *Journal of General Psychology* 37 (1947): 125–8.

22 Grey Walter, *The Living Brain*, London: Duckworth, 1953, 123; Pickering, *The Cybernetic Brain*, 164.

23 Ross Ashby, 'Principles of the Self-Organizing System', in *Principles of Self-Organization*, ed. Heinz von Foerster and George Zopf, New York: Pergamon Press, 1962, 255, 257.

24 Donald Hebb, *The Organisation of Behavior*, New York: Wiley & Sons, 1949. In this book Hebb defined his theory of neuroplasticity as 'connectionist' – a term later repurposed by Frank Rosenblatt to define 'connectionism' as the paradigm of artificial neural networks; see chapter 9.

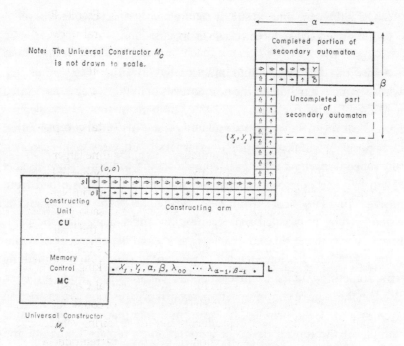

Figure 6.2. Operating arm of the universal constructor. John von Neumann, *Theory of Self-Reproducing Automata* (edited by Arthur Burks), Urbana, Il: University of Illinois Press, 1966, 371.

Hebb recorded a peculiar phenomenon whereby neurons that were simultaneously stimulated also strengthened their connection:

> When one cell repeatedly assists in firing another, the axon of the first cell develops synaptic knobs (or enlarges them if they already exist) in contact with the soma of the second cell... The general idea is an old one, that any two cells or systems of cells that are repeatedly active at the same time will tend to become 'associated' so that activity in one facilitates activity in the other.[25]

Since then, this brain behaviour has been known as the Hebbian theory of 'cell assemblies' and has been conveyed in the famous dictum 'Neurons that fire together, wire together.' Hebb's can be considered as both the first codified rule of neuroplasticity and the first rule of self-organisation for

---

25 Hebb, *Organisation of Behavior*, 63–70.

machine learning algorithms. Cognitive scientist Frank Rosenblatt accordingly conceived the first operative artificial neural network – the perceptron – as a self-organising machine and attempted to implement the Hebbian rule in its functioning (see below and chapter 9).

Theories of self-organisation and the early digital computer

The research field about self-organisation was at the time larger than it is perceived today: it would suffice to remember that sooner or later, all the key pioneers of the digital computer, such as John von Neumann, Konrad Zuse, and Alan Turing, explored self-organisation as a technique of computation. Von Neumann (the designer of the main architecture of digital computers that still to this day bears his name) was investigating radical forms of automation while working for the US military, speculating about a machine – the Universal Constructor – that could reproduce and repair itself (the army was of course interested in applying this idea to self-reproducing and self-repairing vehicles and pieces of artillery). The observation of processes of reproduction in 'living organisms' inspired their simulation in 'computing machines' and established the questionable analogy between organic cells and computational units.[26] The Universal Constructor was one implementation of the general theory of cellular automata – that is, a configuration of computational units that change and evolve like organic cells in a two-dimensional space (see fig. 6.2). In this space, basically, cellular automata are clusters of elements that change configuration and move according to the neighbouring 'cells', composing geometric figures that evolve, claiming to mimic, in this way, life forms in the natural environment. At the Hixon symposium in 1948, Neumann urged the other delegates to understand computation (including Turing machines and artificial neural networks) as a form of self-organisation, arguing that self-reproducing units could perform all standard operations of

---

26 It was Erwin Schrödinger's 1943 lecture 'What Is Life?', in particular, that inspired a link between the negative entropy of photosynthesis and the self-reproduction of life through a code (DNA had not yet been discovered). See also Peter Asaro, 'Heinz von Foerster and the Bio-Computing Movements of the 1960s', in *An Unfinished Revolution? Heinz von Foerster and the Biological Computer Laboratory 1958–1976*, ed. Albert Müller and Karl H. Müller, Vienna: Edition Echoraum, 2007.

computation just by replicating themselves like biological cells.[27] The idea of cellular automata would go on to register a lasting influence. Known for developing the first programmable electric computer in Berlin in 1938, Zuse proposed extending the logic of cellular automata to physics and the general laws of the universe. His 1967 book, *Rechnender Raum* (*Calculating Space*), considered the universe to be composed of discrete spatial units that self-organise as cellular automata – that is, according to the status and behaviour of neighbouring units.[28] According to Zuse, the energy interactions between atoms can be formalised as units of computation and, following this approach, one could rewrite the laws of physics – for instance gravitation – in a combinatorial fashion. In this sense, Zuse saw the theory of calculating space as a paradigm that would supersede quantum psychics in the way the latter had superseded classical physics.[29] From an epistemological point of view, we here encounter once again a mechanical paradigm that openly aspires to become a paradigm of nature, not of biological laws in this case but of physical ones. Precisely, a finite-state machine, such as the digital computer, is elevated to become the ontological model for the structure of the universe itself.

Turing's essay 'The Chemical Basis of Morphogenesis' (published in 1952, two years before his death) also belongs to the tradition of self-organising computation.[30] In this late paper, Turing envisioned the molecules of organisms as self-computing actors that, through their interaction, express complex autopoietic structures. He attempted, with this approach, to model tentacle patterns in hydra, whorl arrangements in plants, gastrulation in embryos, dappling in animal skin, and phyllotaxis in flowers as forms of self-organising computation. To generate such patterns (known since as 'Turing patterns'), he used one of the first mainframe computers of Manchester University, though he also performed a great number of calculations by hand (see fig. 6.3). Turing warned that

---

27 See also Alex Galloway, 'Creative Evolution: Nils Aall Barricelli's Mathematical Organisms', *Cabinet* 42 (Summer 2011).

28 Konrad Zuse, 'Rechnender Raum', in *Elektronische Datenverarbeitung* 8 (1967), translated as *Calculating Space*, Cambridge, MA: MIT Technical Translation, 1970.

29 Zuse, *Calculating Space*, 5.

30 Alan Turing, 'The Chemical Basis of Morphogenesis', *Philosophical Transactions of the Royal Society B.* 237, no. 641 (1952).

The Self-Organisation of the Cybernetic Mind

> 'this model will be a simplification and an idealization, and consequently a falsification', though he expressed, as any cybernetician would, the hope that 'the imaginary biological systems which have been treated, and the principles which have been discussed, should be of some help in interpreting real biological forms'.

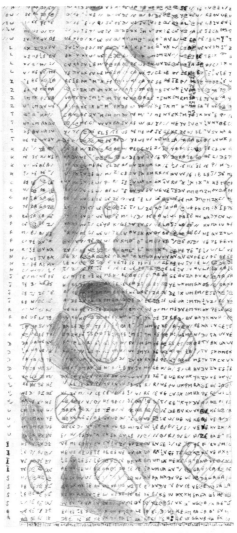

Figure 6.3. Diagram showing patterns of dappling and calculations. Alan Turing, ca. 1950. Sheet AMT/K3/8, Turing Archive, King's College Cambridge, particular.

Military concerns about self-organising networks

Self-organisation theories belonged not only to cybernetic dreams of living automata, but also to the armamentarium of Cold War rationality.[31] As often, the main sponsor of research in this field was the US military, which expressed interest in the logic of self-organisation as an alternative, more efficient, means of computation. At the end of the 1950s, the US Office of Naval Research (ONR) decided to sponsor a series of symposia on self-organisation that provide a historical documentation of the wide reception of artificial neural network research at the time. Strangely, even Margaret Boden's monumental history of AI failed to register the existence of these other symposia, with the result that historical importance is repeatedly granted only to the 1956 Dartmouth workshop on AI.

In May 1959, Marshall Yovits, head of the newly established Information System Branch at the ONR, chaired the conference on 'Self-Organising Systems' in collaboration with the Illinois Institute for Technology. Yovits invited cyberneticians from both the connectionist and symbolic AI camps.[32] Somehow anticipating the forthcoming confrontation of these two paradigms and the limits of the latter, he argued that

> certain types of problems, mostly those involving inherently non-numerical types of information, can be solved efficiently only with the use of machines exhibiting a high degree of learning or self-organizing capability. Examples of problems of this type include automatic print reading, speech recognition, pattern recognition, automatic language translation, information retrieval, and control of large and complex systems. Efficient solutions to problems of these types will probably require some combination of a fixed stored program computer and a self-organizing machine.

---

31 See Paul Erickson et al., *How Reason Almost Lost Its Mind: The Strange Career of Cold War Rationality*, Chicago: University of Chicago Press, 2013.

32 The conference was also attended by the representatives of symbolic AI – Allen Newell, Cliff Shaw, and Herbert Simon. See Stephanie Dick, 'Of Models and Machines: Implementing Bounded Rationality', *Isis* 106, no. 3 (2015): 623–34. See also Shunryu Garvey, 'The "General Problem Solver" Does Not Exist: Mortimer Taube and the Art of AI Criticism', *IEEE Annals of the History of Computing* (2021): 60–73.

The conference exemplified the interdisciplinary ambitions of cybernetics, with Yovits highlighting how researchers from the fields of life sciences such as psychologists, embryologists, and neurophysiologists were working together to comprehend the characteristics of self-organising biological systems, while on the other hand, 'mathematicians, engineers, and physical scientists were attempting to design artificial systems which could exhibit self-organizing properties'.[33]

The proceedings of the conference help cast a light on the debate on self-organisation beyond the canonical themes of cybernetics. Despite the variety of positions on display, the main focus of the conference throughout remained the self-organisation of computing networks. The electrical engineer Belmont Farley opened the conference with an overview of the main visual 'systems which automatically organize themselves to classify environmental inputs into recognizable percepts or patterns'. Farley's paper was the continuation of his studies carried out during World War II, when, during the bombing of London, he had been responsible for testing a new type of radar against the low-flying Luftwaffe, showing how the self-organisation of the visual field was clearly already a concern of military automation. Other contributors of the conference included zoologist Robert Auerbach, who attempted to describe 'the organisation and reorganisation of embryonic cells' in mathematical terms – specifically as a 'transfer of information (induction)' in living cells that were dubbed as 'growing automata'. In this extended research project to grasp the principle of self-organisation in nature, the British cybernetician Gordon Pask contributed the idea of initiating a 'natural history of networks' with the intention of proving a similarity of qualities between social and natural ones:

> [If] an observer wishes to use any self-organizing potentialities the network may have, then he must look at the network as though he were a natural historian ... using the term 'network' in a general sense, to imply any set of interconnected and measurably active physical entities. Naturally occurring networks, of interest because they

---

33 Marshall Yovits and Scott Cameron (eds), *Self-Organizing Systems*, New York: Pergamon Press, 1960, v–vi.

have a self-organizing character, are, for example, a marsh, a colony of micro-organisms, a research team, and a man.[34]

This 1959 symposium is important also to see, in perspective, McCulloch and Pitts's original idea of artificial neural networks, which were not just input–output black-boxed machines but systems seeking to imitate and embody neuroplasticity. A decade after his and Pitts's founding 1943 paper, McCulloch took part in the symposium to stress that self-organisation is key to neural networks and that the same principle should be used in the design of an 'infallible network of fallible neurons'. McCulloch's intuition (which is still today both the strength and the limit of machine learning) was that computation does not need to be accurate to be efficient but can instead be based 'on redundancy of calculation'. In the paradigm of artificial neural networks, 'information is brought to a lot of so-called neurons, and these crummy neurons, working in parallel computation, can come out with the right answer even though the component neurons are misbehaving'.[35]

A following event sponsored by the ONR was the 'Symposium on Principles of Self-Organization', convened by Heinz von Foerster in 1960 in collaboration with the Biological Computer Laboratory of the University of Illinois at Urbana-Champaign. The proceedings of this symposium also testify that McCulloch and Pitts's, as well as Rosenblatt's, artificial neural networks were part of a larger archipelago of similar research projects on self-organisation.[36] Most of the contributions were related to the fields of neural networks (also referred to as 'random networks') and covered issues such as learning techniques, error correction, inductive inference, distributed memory, pattern recognition, as well as prototypes of self-organising hardware units, or neuristors (what today would be called neuromorphic chips).

Among the participants should be noted the presence of the neoliberal economist Friedrich Hayek, who in 1952 authored an oft-overlooked

---

34 Gordon Pask, 'Natural History of Networks', in ibid., 232.

35 For McCulloch the brain had to be treated as an organisation of information abstracted from any energetic and material dimension: Warren McCulloch, 'The Reliability of Biological Systems', in Yovits and Cameron, *Self-Organizing Systems*, 265.

36 Heinz von Foerster and George Zopf (eds), *Principles of Self-Organization*, New York: Pergamon Press, 1962.

treatise on connectionism, *The Sensory Order* (see chapter 8). Hayek's presence signals the overlapping interests of the military, economists, cyberneticians, and industrialists regarding the subject of self-organisation. Cyberneticians, for their part, were eager to prove their theories for the benefit of the economy and industry. Not by chance was the symposium opened by Stafford Beer's 'electroencephalogram of one of Britain's largest steel mills', which considered the organisation of a factory as equivalent to a brain.[37] Beer had already proposed a 'sketch of a cybernetic factory' in his 1959 book *Cybernetics and Management*, which exemplified the political attitude of cybernetics to shape machines, organisms, and workers all in like manner. In spite of his later collaboration with the socialist government of Salvador Allende for the project Cybersyn in 1971, Beer maintained a managerial view of the economy. At this conference, his primary concern appears to be the self-organisation of industrial management – the eye of the master – rather than any other aspect of society.

The end of the Cybersyn project is exemplary of the twisted fate of the politics of self-organisation. Cybersyn was a communication network for the management of the Chilean economy. It was contemporaneous with Arpanet (the progenitor of the internet funded by the US Department of Defense), yet less advanced. Arpanet featured a decentralised network based on packet-switching communication, while Cybersyn remained a centralised web of teletypes linked to a single mainframe computer. Arpanet was based on the idea that a decentralised communications network could survive an enemy attack, as the brain's neural networks reorganise themselves in case of injury. The US Army co-opted this idea of network plasticity before anyone else. The Cybersyn project was terminated when a CIA-backed coup d'état brought Salvador Allende's life (and Chilean democracy) to an end.[38]

---

37 Ibid., ix.

38 See Andrew Pickering, 'The Science of the Unknowable: Stafford Beer's Cybernetic Informatics', *Kybernetes* (2004); Eden Medina. *Cybernetic Revolutionaries: Technology and Politics in Allende's Chile*, Cambridge, MA: MIT Press, 2014.

From linear to self-organising information

From the vantage point of today's deep learning, it is the participation of Frank Rosenblatt in these conferences that appears of key significance. In 1957, at the Cornell Aeronautical Laboratory in Buffalo, New York, Rosenblatt developed the first statistical artificial neural network 'perceptron', which, after several generations of improvements, would ground the deep learning architecture. Rosenblatt attended these conferences to defend his prototypes, which were quite fragile experiments. Emerging from diverse theoretical and technical influences (see chapter 9), the perceptron was conceived as a self-organising computing network that, in order to recognise a pattern, would find the optimal value of its parameters by gradually adjusting them to the input data. As Rosenblatt remarked in one of these conferences, the perceptron 'arrives at its organisation spontaneously, rather than having it built into the system'.

What did the perceptron look like as a device? One of the first prototypes, the Mark I Perceptron, was an analogue-digital machine comprising an input device of 20 × 20 photocells (called 'retina') connected through wires to a layer of artificial neurons that resolved into a single output (a light bulb turning on or off, to signify if a pattern was recognised or not). The retina of the perceptron recorded simple shapes such as letters and triangles, passing electrical signals to a series of artificial neurons that would sum them up and memorise a result according to a cumulative logic – somehow implementing Hebb's rule to form cell assemblies, as Rosenblatt initially intended.

At the 1959 conference 'Self-Organising Systems', Rosenblatt's contribution was concerned with the generalisation of visual stimuli – that is, with the recognition of similar patterns in a noisy environment. His paper aimed at explaining 'how a brain, or brain-like system, can recognize similarity among the various possible transformations of a sensory pattern, or image'. The problem he addressed was the capacity of a statistical neural network for pattern recognition to generalise beyond the cases of its training dataset. In simple terms, it was 'the dilemma of distinguishing the [letter] N from the Z' under different visual orientations. The way Rosenblatt illustrated his project could still be used today to illustrate the working of a deep learning algorithm:

This system is capable of 'abstracting' those transformations which are most common in a particular environment, and applying them to new stimuli, which may be quite different in form from any which it has seen. It seems to accomplish all of the results of more rigidly designed systems, but arrives at its organisation spontaneously, rather than having it built into the system. It is actually a system which 'learns to learn', in the sense that prior to the preconditioning experience it would be able to generalize from a given stimulus to its transform only by the slow and laborious method of contiguity generalization, while after having seen a suitable preconditioning sequence (not including the stimuli to be used for test purposes), it performs the same task directly and without the requirement of any appreciable learning period.[39]

As Rosenblatt explained at the 1961 conference 'Principles of Self-Organisation', the design of the perceptron was different from previous artificial neural networks precisely due to its self-organising behaviour, which was possible via its 'reinforcement control system'.[40] Although it was a 'brain model' according to Rosenblatt, the perceptron was far more about the self-organisation of information than the mimicry of organic structures. It marked, therefore, not so much a biomorphic turn in computation as a topological one. Computer scientists Marvin Minsky and Seymour Papert renamed connectionism, somewhat pejoratively, as 'computational geometry' because it was based on the calculation of spatial relations rather than being an instance of 'true AI'.

This topological turn marked, more generally, the passage from a paradigm of linear information to one of self-organisation under which, I argue, the large family of machine learning techniques should be considered. Indeed, it introduced a second spatial dimension into a model of computation that, until then, had been understood primarily within the linear dimension of numerical computers. Instead of processing a visual matrix via a top–down algorithm (as in a traditional program

---

39 Frank Rosenblatt, 'Perceptual Generalization over Transformation Groups', in Yovits and Cameron, *Self-Organizing Systems*, 95.

40 By comparison, Oliver Selfridge's Pandemonium network relied on handcrafted nodes to extract predetermined features. Oliver G. Selfridge, 'Pandemonium: A Paradigm for Learning', *Proceedings of the Symposium on the Mechanization of Thought Processes* 1 (1959): 511–29.

of instructions, following the scheme of the Turing machine), the perceptron computed the pixels of its visual matrix in a bottom–up and parallel fashion according to their spatial disposition. With respect to computational forms, artificial neural networks such as the perceptron implicitly marked the bifurcation between these two paradigms: that of *linear information* (broadly represented by media such as telegraphs and numerical computers, as well as symbolic AI) and that of *self-organising information* (represented by cybernetic systems, cellular automata, and, ultimately, connectionist AI). Historians of AI Hubert and Stuart Dreyfus summarised the epistemic distinction between the symbolic and connectionist schools in a similar way:

> One faction saw computers as a system for manipulating mental symbols; the other, as a medium for modeling the brain. One sought to use computers to instantiate a formal representation of the world; the other, to simulate the interactions of neurons. One took problem solving as its paradigm of intelligence; the other, learning. One utilized logic; the other, statistics. One school was the heir to the rationalist, reductionist tradition in philosophy; the other viewed itself as idealized, holistic neuroscience.

This book does not recapitulate the saga of linear information in the twentieth century – of cybernetic feedback loops, sequential media, and symbolic AI – but tells the parallel story of self-organising information which is necessary, however, to emancipate from the legacy of biomorphism and research paradigms such as Artificial Life.[41]

Computational thinking and mechanistic analogies of the mind

There exists a considerable misunderstanding about cybernetics' scientific aspirations. In reality, cybernetics was not a science but a school of engineering in drag – one with sufficient self-confidence to extend its informational and computational analogies to several aspects of nature and society. This book tries to clarify that, rather than designing

---

41 Hubert Dreyfus and Stuart Dreyfus, 'Making a Mind versus Modeling the Brain: Artificial Intelligence Back at a Branchpoint', *Daedalus* 117 (1988): 15–44.

machines like organisms (*biomorphism*) as they professed, cyberneticians ultimately envisioned organisms like machines (*technomorphism*), which were mirroring their own surrounding social order (*sociomorphism*). Like the philosophies of nature from earlier centuries (the canonical example being La Mettrie's *L'homme Machine* of 1747), cyberneticians projected on the ontology of nature and the brain the technical composition of their time, made up of telegraph networks, electromechanical relays, feedback systems, and television scanners. Cyberneticians did not pursue a scientific and experimental but rather a speculative (and often naive) method of *analogy*, mapping preconstituted rules onto nature rather than making hypotheses about new ones. McCulloch, Pitts, and von Neumann's insistence that brain neurons are 'switching organs' functionally equivalent to electromechanical relays is a good example of cybernetics' presumptuous analogies.[42]

The analogy between organisms and machines appears, at first glance, to be an issue of epistemic translation between the disciplines of engineering and biology, but, in fact, it points to a more profound attitude of cybernetic engineering: What are the ethical implications of seeing an industrial machine as an organism, a living being? As much as 'computer science', cybernetics was not a science but an artificial language, a manual of instructions for machine components – a 'machine semiotics' which happened to be forcibly translated into an ontology of nature.[43] But, if it is true that technology can influence scientific paradigms and models of the universe, it is equally true that technology has its own demons and is shaped by external forces within its own domain. Communications networks such as the telegraph, for instance, are not simply technical apparatuses but social institutions. Cyberneticians

---

42 'The neuron, as well as the vacuum tube . . . are then two instances of the same generic entity, which it is customary to call a *switching organ* or *relay organ* . . . The basic switching organs of the living organisms, at least to the extent to which we are considering them here, are the neurons. The basic switching organs of the recent types of computing machines are vacuum tubes; in older ones they were wholly or partially electromechanical relays.' John von Neumann, 'The General and Logical Theory of Automata', in Jeffress, *Hixon Symposium*, 12.

43 For the expression 'machine semiotics' see chapter 2 of this book and Simon Schaffer, 'Babbage's Calculating Engines and the Factory System', *Réseaux* 4, no. 2 (1996): 280. Herbert Simon tried to define 'the sciences of the artificial' in the homonymous book, but the result is far from rigorous. Herbert Simon, *The Sciences of the Artificial*, Cambridge, MA: MIT Press, 1969.

projected the technical composition that was implied in their own profession, in the form of their labour and knowledge, onto a new paradigm of the world. Specifically, they projected onto nature forms of self-organisation that were already part of the division of labour and technical organisation of their surrounding society. The way cyberneticians claimed to imitate the self-organisation of living beings to build machines implicitly revealed more about the organisation of society and labour relations of their age than about nature.

McCulloch once claimed that 'every robot suggests a mechanistic hypothesis concerning man'.[44] This thesis of cybernetic epistemology argues that the invention of machines may help discover insights about the workings of the human, following the reductionist idea that machines and organisms exist in the same universe and thus must obey the same physical rules. But the word 'robot' is here revealing, because of its industrial and feudal legacy, and it can suggest another meaning. Interpreted in its full implications, this thesis may actually imply that every form of labour automation says something about the cognitive models of a given age. At the end, it confirms one of the main concerns of historical epistemology: that the organisation of labour in a given epoch influences the formation of technologies and instruments, and thereafter of scientific paradigms, conceptions of nature, and models of the mind too.

What has been illustrated for the industrial age appears to be true also for the information age: the *means of production* (not simply telegraphs and computers, in this case, but also artificial neural networks) imitate – in their inner design – the *relations of production*, that is the extended organisation of labour in society. Information technologies increased their hold over society by this adaptation, not by the power of a technological a priori (as techno-determinists maintain), but through a social a priori – that is, by their inborn capacity to capture social cooperation. The nineteenth-century *labour theory of automation* finds confirmation also in the information age.

Eventually, it comes as no surprise that the most successful AI technique, namely artificial neural networks, is the one that can best mirror,

---

44 Warren McCulloch to Hans Lukas Teuber, 10 December 1947. Quoted in Ronald R. Kline, *The Cybernetics Moment: Or Why We Call Our Age the Information Age*, Baltimore: Johns Hopkins University Press, 2015, 46. The term 'robot' derives from the feudal-era Czech word for 'servant'.

and therefore best capture, social cooperation. The paradigm of connectionist AI did not win out over symbolic AI because the former is 'smarter' or better able to mimic brain structures, but rather because inductive and statistical algorithms are more efficient at capturing the logic of social cooperation than deductive ones. By tracing the evolution from linear to self-organising information, the history of data analytics, machine learning, and AI can begin to be seen in perspective as a grand process of self-organisation within the technosphere to follow the transformation of the social order.

The disciplines and denominations of information theory, cybernetics, artificial intelligence, and computer science all consolidated in the 1950s.[45] While the US military, as we have seen, played an important role in funding many of these projects, it would be mistaken to presume they deserve sole credit for the origination of these disciplines. Indeed, this book looks to widen this established genealogy. Against technodeterminist and strong internalist readings of information technologies, I have proposed an equally strong externalist hypothesis: that the design of information machines responded – even at the level of the logical forms of their algorithms – to the forms of social interaction at large.[46] In the twentieth century, in other words, it was not information technologies that primarily reshaped society, as the mythologised vision of the 'information society' implies; rather, it was social relations that forged communication networks, information technologies, and cybernetic theories from within. Information algorithms were designed according to the logic of self-organisation to better capture a social and economic field undergoing radical transformation.

---

45 Respectively, information theory was introduced by Shannon in 1948, cybernetics by Wiener in 1948, artificial intelligence by McCarthy in 1955, machine learning by Arthur Samuel in 1959, and computer science by George Forsythe only in 1961. See George Forsythe, 'Engineering Students Must Learn Both Computing and Mathematics', *Journal of Engineering Education* 52 (1961); and, in particular, Donald Knuth, 'George Forsythe and the Development of Computer Science', *Communications of the ACM* 15 (1972). For the other references see above.

46 For an opposite reading, see Friedrich Kittler's thesis that World War II was the main driving force behind the development of the digital computer. See Geoffrey Winthrop-Young, 'Drill and Distraction in the Yellow Submarine: On the Dominance of War in Friedrich Kittler's Media Theory', *Critical Inquiry* 28, no. 4 (2002): 825–54.

Autonomy and automation

In the second half of the twentieth century, 'Autonomy!' emerged as the common slogan for both cybernetics and the emerging counterculture. High-level cyberneticians funded by the US army were discussing 'principles of self-organisation' in organisms and machines just as anti-authoritarian movements were proposing the same for social and political institutions. As such, these two tendencies debated, for different purposes, the ability of a system to give itself new rules over and against an external ruler (which is, in fact, the original meaning of 'autonomy'). They were both, each in their own respect, forms of political avant-garde and a response to the dominion of outdated regimes: European fascism, Stalinist totalitarianism, and American capitalism. The terms 'cybernetics' and 'beat generation' were both coined, coincidentally, in 1948. A few years later, Norbert Wiener defined both fascism and Western corporatism as ideologies of 'the inhuman use of human beings' against which cybernetics purported to offer a more 'human use of human beings'.[47] But, where cybernetics in fact bolstered US military primacy during and after World War II, the counterculture and the student movement firmly boycotted the Vietnam War and the arms race. The project of autonomy obviously meant different things to these different parties. For anti-authoritarian movements, it represented the freedom of self-determination and a means to constitute new institutions and alternative forms of life. For the cyberneticians, it was the technological utopia of full automation and enlightened societal control: a military and industrial fantasy which also included the project of AI. That even the military – that most traditionally hierarchical structure – also had a vested interest in forms of distributed communication and self-organising networks is a sign of deeper transformations.

In the 1960s, the Free Speech Movement at the University of California, Berkeley, rightly condemned the first mainframe computers as technologies of war and social control in the hands of government and corporations. Media scholar Fred Turner remembers when, on 2 December 1964, in front of more than five thousand students at the University of California, Berkeley, activist Mario Savio delivered an

---

47 Norbert Weiner, *The Human Use of Human Beings*, Boston: Houghton Mifflin, 1950, 71.

incendiary speech in which he 'uttered three sentences that would come to define not only the Free Speech Movement at Berkeley, but the countercultural militancy of the 1960s across America and much of Europe as well':

> There's a time when the operation of the machine becomes so odious, makes you so sick at heart, that you can't take part, you can't even tacitly take part. And you've got to put your bodies upon the gears and upon the wheels, upon the levers, upon all the apparatus, and you've got to make it stop. And you've got to indicate to the people who run it, to the people who own it, that unless you're free, the machine will be prevented from working at all.[48]

For Turner, Savio's speech evoked memories of the pre-digital era, with images of workers physically wrestling with machines on factory floors. However, he also linked the term 'machine' to the modern society's dependence on information technology, which was beginning to significantly organise social relations as well.[49]

Such criticism of information technologies changed polarity in the following decade: computer science absorbed the aspirations of the earlier counterculture, while the counterculture itself claimed the emancipatory potential of information technologies (and eventually mutated into the so-called cyberculture). The controversial imbrication of social autonomy and technological automation was already present, albeit underground, in the debates of the 1960s. Components of the counterculture, especially those inspired by Eastern spirituality, developed a naive attraction to cybernetics.[50] The *Whole Earth Catalog*, published in California between 1968 and 1972, came to represent a culmination and synthesis of both cybernetic and ecological traditions. Richard Brautigan photographed this convergence in his famous satirical poem 'All Watched Over by Machines of Loving Grace'.[51] On the other hand,

---

48 Fred Turner, *From Counterculture to Cyberculture: Stewart Brand, the Whole Earth Network, and the Rise of Digital Utopianism*, Chicago: University of Chicago Press, 2010, 11.
49 Ibid., 11.
50 See Pickering, *The Cybernetic Brain*.
51 Richard Brautigan, *All Watched Over by Machines of Loving Grace*, San Francisco: Communication Company, 1967.

European voices such as that of Herbert Marcuse from the Frankfurt school and the autonomist Marxists reclaimed automation in the battle of emancipation from industrial labour. In Italy, a famous slogan of the *autonomia* cried: 'Lavoro zero e reddito intero, tutta la produzione all'automazione' (Zero work and full income, all production to automation).

The terms 'autonomy', 'autonomous', 'automation', and the more ambivalent 'autonomisation' (meaning, depending on the context, 'being automated' or 'becoming autonomous') are not equivalent and also differ from 'self-organisation'. Etymologically, the classical Greek term *autonomia* – from *autos* ('self') and *nomos* ('law') – signifies the power to give oneself new habits, rules, and laws. Modernity recognises this power as one belonging to legislative institutions, especially to the constituent assembly that founds the political order of the state.[52] Autonomy is simultaneously a constituent and destituent power: each time a new rule is invented, an old one can be subverted, nullified, or incorporated by the new invention. But the opposite is also true: any time a rule is broken, an anomaly takes form and a new constitution – a new vision of the world – is implied.

In cybernetics, autonomy was defined as the capacity of a technical system of multiple agents to find a new organisation and equilibrium in relation to external inputs – namely, the capacity to adapt to the environment. In this way, a technical system was said to show emergent properties that might be perceived as 'intelligent' by a human observer. These questions continue to haunt the dream of artificial intelligence, even now: Can a finite-state automaton – that is, a computer – show properties of autonomy? Can a computer programmed to follow strict rules rebel against its core instructions and invent new ones? If autonomy is the power to invent a new rule, automation can be defined as the blind following of a rule, as is the case with computation. In this regard, the Austrian philosopher Ludwig Wittgenstein remarked that 'following a rule' will always have a different meaning for a human and a machine.[53]

---

52 In early research, the German system theorist Niklas Luhmann noticed that the institutional order had to face the growing normative role of technical apparatuses: Niklas Luhmann, *Recht und Automation in der öffentlichen Verwaltung: Eine verwaltungswissenschaftliche Untersuchung*, Berlin: Duncker & Humblot, 1966.

53 See Stuart G. Shanker, *Wittgenstein's Remarks on the Foundations of AI*, London: Routledge, 1998.

Considering the recent debates on AI bias as well as the speculation on the risks of 'superintelligence', one wonders whether the game of AI is still being played within the domain of automation (following a rule) rather than the domain of autonomy (that of breaking rules).

To conclude, a competing claim: technologies of automation have always been responses to social autonomy, and cybernetic techniques of self-organisation such as artificial neural networks, similarly, have been avatars of the emergent social relations of their day. In hindsight, both cybernetics and the post-war social movements were directly related to the autonomisation of knowledge and information in labour processes and social behaviours, which had triggered the rise of new media and technologies. These points have over the years become a conventional interpretation in theories of knowledge society and information economy, to the point that even neoliberal economic paradigms, such as Hayek's spontaneous order of markets, or 'catallaxy', can be considered responses to the increased exchange of information in society at large (see chapter 8). Materialist historians concede the dialectical relations of the two movements – between the drive for social autonomy by new generations of workers, on one hand, and the appearance of new technologies of automation, on the other. Ultimately, the diverse projects of automation after World War II were a way to govern developing social forces – that is, to organise a 'control revolution' (as Beniger defined it) against a more rebellious society.[54] It is not by chance that, at least in the Global North, students and computer programmers were transformed into a new political subjectivity similar to the industrial workers' movement, given a global economy more and more dependent on information, knowledge, and science as key economic drivers.

In the late 1960s, political philosopher Mario Tronti proposed to reverse a thesis which was then mainstream also in Marxism: capitalist development was always considered to shape workers' organisation and their politics. To the contrary, Tronti claimed that capitalist development, including technological innovation, was always triggered by workers' struggles. Interestingly, for a European intellectual such as Tronti, 'the working-class struggle reached its highest level of development between 1933 and 1947, and specifically in the United States',

---

54 James Beniger, *The Control Revolution: Technological and Economic Origins of the Information Society*, Cambridge, MA: Harvard University Press, 1986.

which are coincidentally the years that witness the rise of cybernetics and digital computation.[55] Radical and unconventional perspectives like this should be explored to narrate the combined evolution of society and technology throughout the twentieth and twenty-first centuries. Across the historical transformations that this book attempts to analyse, it appears that the project of AI has never been truly *biomorphic* (aiming to imitate natural intelligence, as mentioned earlier) but implicitly *sociomorphic* – aiming to encode the forms of social cooperation and collective intelligence in order to control them.[56] The destiny of the automation of intelligence cannot be seen as separate from the political drive to autonomy: it was ultimately the self-organisation of the social mind that gave form and momentum to the project of AI.[57]

---

55 Mario Tronti, 'Postscript of Problems', in *Operai e capitale*, 2nd ed., Torino: Einaudi, 1971, translated as *Workers and Capital*, London: Verso, 2019, 294. See also Raniero Panzieri, 'Sull'uso capitalistico delle macchine nel neocapitalismo', *Quaderni Rossi* 1 (1961): 53–72.

56 See also Matteo Pasquinelli, 'Abnormal Encephalization in the Age of Machine Learning', *e-flux* 75 (September 2016).

57 For an overview of the ideas of the social brain and collective intelligence in contemporary political theory, see Charles Wolfe, 'The Social Brain: A Spinozist Reconstruction', *ASCS*, Proceedings of the Ninth Conference of the Australasian Society for Cognitive Science, 2009, 366–74.

7

The Automation of Pattern Recognition

> *As the industrial revolution concludes in bigger and better bombs, an intellectual revolution opens with bigger and better robots.*
>
> Warren McCulloch, Hixon symposium, 1948[1]

> *A new, essentially logical, theory is called for in order to understand high-complication automata and, in particular, the central nervous system. It may be, however, that in this process logic will have to undergo a pseudomorphosis to neurology to a much greater extent than the reverse.*
>
> John von Neumann, Hixon symposium, 1948[2]

The controversy about Gestalt perception

At the core of the intuition that paved the way for artificial neural networks lies an enduring controversy: whether or not human perception is an act of cognition that can be analytically represented and therefore mechanised. The confrontation about this issue flared up in the 1940s during the Macy conferences between the cyberneticians (who

---

1 Lloyd A. Jeffress (ed.), *Cerebral Mechanisms in Behavior: The Hixon Symposium*, New York: Wiley, 1951, 42.
2 Ibid., 24.

argued that the perceptual field as a whole can be computed by machines, such as simple electric relays) and the Gestalt school (who maintained that a machine would never be able to emulate the complex synthetic faculty of the human mind).[3] It constituted a new chapter in the ongoing Gestalt controversy from Europe; yet this time, exiled from Nazi Germany to the United States, its protagonists included mathematicians and engineers in addition to the previous cast of philosophers, psychologists, and neurologists.[4] It was in the aftermath of this debate, in fact, that the expression 'Gestalt perception' gradually morphed, in military reports and academic publications, into the more familiar 'pattern recognition' and gave momentum to the experiments with artificial neural networks.[5]

Cyberneticians such as Warren McCulloch and Walter Pitts proposed a dramatic simplification of the act of perception, at which any art historian would baulk. However, their research provoked a breakthrough in the 'automation of perception' that would unfold, half a century later, in the well-known exploits of deep learning.[6] Before cyberneticians reduced the faculty of perception to a simple binary act of classification (whether or not an image belongs to a given class), Gestalt scholars such as Max Wertheimer, Kurt Koffka, and Wolfgang Köhler had advanced a more complex theory of perception. Gestalt theory famously referred to the fact that one perceives the whole before its parts – a melody, for instance, remains discernible even when it is played in a different scale

---

3 See Claus Pias (ed.), *Cybernetics: The Macy Conferences 1946–1953; The Complete Transactions*, Berlin: diaphanes, 2016; Steve Heims, *The Cybernetics Group*, Cambridge, MA: MIT Press, 1991, chaps. 9–10. Steve Heims called it the 'Gestalten go to bits' controversy.

4 See Mitchell G. Ash, *Gestalt Psychology in German Culture, 1890–1967: Holism and the Quest for Objectivity*, Cambridge: Cambridge University Press, 1998.

5 The German term 'Gestalt' is often translated in English as 'form', although German possesses the distinctive word *Form*. The best English translation is probably 'configuration', which is a form understood in its general structure and internal implications. The term 'pattern recognition' was popularised by *Scientific American* in 1960: Oliver G. Selfridge and Ulric Neisser, 'Pattern Recognition by Machine', *Scientific American* 203, no. 2 (1960): 60–9. See also William D. Rowe, 'Gestalt Pattern Recognition with Arrays of Predetermined Neural Functions', in *International Joint Conference on Artificial Intelligence* (1969): 117–40.

6 Paul Virilio used the expressions 'automation of perception' and 'artificial vision' in relation to the neural network perceptron in *The Vision Machine*, Indiana University Press, 1994. See chapter 9 of this book.

and with different sounds. It pointed to diverse phenomena of image closure, such as the subconscious perception of a configuration in its totality in spite of limited information about its elements (e.g., the perception of a triangle in the abstract relation of three separated elements). Eventually, it was formalised in the principles of *Prägnanz* (proximity, similarity, continuity, connectedness, etc.) and encapsulated in the famous motto 'The whole is other than the sum of its parts'.

Textbooks on machine learning repeat the standard account that McCulloch and Pitts's invention of artificial neural networks was inspired by the neurophysiology of the brain, while overlooking the broader intellectual context that includes their confrontation with the Gestalt school. Though artificial neural networks may be regarded as a brilliant solution to a challenge of automation, it was the Gestalt school that established the perception of a visual form as a key issue to address. The specific problem that Gestalt theory urged cyberneticians to resolve through a computational architecture was the recognition of a pattern, such as an image. Is the recognition of an image equivalent to logical reasoning? McCulloch and Pitts thought so, therefore arguing that any visual pattern (even a complex, fuzzy, and incomplete one) can be fully defined by computing the relation of its elements within the visual field. In their view, the recognition (or classification) of a pattern could be resolved by an algorithm to compute a large input into a simple binary output (0 or 1), representing in this way a simple binary question: 'Does the image belong to a given class or not?'

After short skirmishes during the Macy conferences, the Gestalt controversy convened at a dedicated event: the Hixon symposium on 'Cerebral Mechanisms and Behavior', which took place at the California Institute of Technology, Pasadena, in 1948, and was attended by McCulloch, von Neumann, and Köhler, as well as famed psychologists Heinrich Klüver and Karl Lashley, among others.[7] The Hixon symposium was a watershed event in the history of computation: even John McCarthy admitted that his felicitous term 'artificial intelligence', which was coined for the Dartmouth conference in 1956, had been inspired by his attendance at this earlier symposium as a graduate student in mathematics.[8]

---

7 For the whole proceedings, see Jeffress, *Hixon Symposium*.
8 Nils John Nilsson, *The Quest for Artificial Intelligence*, Cambridge: Cambridge University Press, 2009, 52. McCarthy only cultivated an interest in propositional logic.

At the time, neither of the opposing factions won the controversy. As physicist and science historian Steve Heims noted, 'In a sense, both were romantics, Köhler in his holism and McCulloch in his mechanism.'[9] As a matter of fact, it was von Neumann who proposed a synthesis of Gestalt theories and computational logic that eventually progressed the field and inspired Rosenblatt's neural network perceptron, which implemented statistical calculus in place of McCulloch and Pitts's rigid logic. This chapter aims to re-examine the influence of Gestalt theory upon the history of AI, and to highlight, in particular, the transformation of artificial neural networks from mediums of logical reasoning into instruments for pattern recognition. The Gestalt controversy and the Hixon symposium are proposed, then, as an alternative history of AI that does not relitigate the (failed) enterprise of symbolic AI and its ambitions to mechanise deductive logic, but shifts the focus to the lineage of connectionism and the automation of inductive logic.

The Gestalt controversy is a cognitive fossil of unresolved problems that is today buried at the core of deep learning, and its study can help us to understand the logical form and limits that twenty-first-century AI has inherited.[10] The Gestalt controversy also points to an *optical unconscious* still operative in contemporary machine learning, as the technique of pattern recognition has been extended from visual to non-visual information datasets over the past few decades. In a peculiar twist of fate, it is the mechanisation of perception as pattern recognition that has come to be traded as the mechanisation of cognition, or artificial intelligence. Nevertheless, despite its origins in the automation of vision, the use of anthropomorphic metaphors of perception to describe the

---

In 1979, he was still keen to clarify that 'machines as simple as thermostats can be said to have beliefs'. John McCarthy, 'Ascribing Mental Qualities to Machines', in *Philosophical Perspectives in Artificial Intelligence*, ed. Martin Ringle, Atlantic Highlands, NJ: Humanities Press, 1979.

9   Heims, *Cybernetics Group*, 241.

10 Google researcher François Chollet has designed the Abstract and Reasoning Corpus (ARC) as a visual test that is deemed simple for humans but still difficult even for current large AI models to solve. Without acknowledging it, ARC covers tasks such as image completion with limited information that had previously been canonised by the Gestalt school as part of their *Prägnanz* principles. See François Chollet, 'On the Measure of Intelligence', preprint (2019), arxiv.org. See also Hannes Bajohr, 'The Gestalt of AI: Beyond the Holism-Atomism Divide', *Interface Critique* 3 (2021): 13–35.

operations of artificial neural networks, as well as today's deep learning, is misleading. As is often repeated, machine vision 'sees' nothing: what an algorithm 'sees' – that is, calculates – are topological relations among numerical values of a two-dimensional matrix. Ultimately, the breakthrough brought about by artificial neural networks was not so much biomorphic as, in this case, *topological*. In other words, it was not so much about imitating the structure of neural networks in the eye's retina as, essentially, about developing techniques of self-organising information to read the visual matrix.

The invention of artificial neural networks

As outlined in the previous chapter, the invention of artificial neural networks was canonised in a milestone paper by McCulloch and Pitts: 'A Logical Calculus of the Ideas Immanent in Nervous Activity' (1943). This was followed by another key text that was a direct response to the Gestalt controversy: 'How We Know Universals' (1947).[11] While the former introduced the idea of a network of artificial neurons to automate logical reasoning, the latter advanced its application to 'the perception of auditory and visual forms' (see fig. 7.1). The passage from the former to the latter marks a logical breakaway. Whereas the 1943 paper proposed neural networks as *deductive machines* for propositional calculus, the 1947 paper pointed towards *inductive machines* for automating pattern recognition. It is worth bearing in mind that McCulloch and Pitts's 1943 paper did not mention computers, because they were not then an established technology.[12]

---

11 On the theoretical background of McCulloch and Pitts, see Tara Abraham, '(Physio)logical Circuits: The Intellectual Origins of the McCulloch–Pitts Neural Networks', *Journal of the History of Behavioural Science* 38, no. 1 (2002): 3–25.

12 At the time of the Gestalt controversy, the digital computer was still an experimental idea and not a technical convention taken for granted. Statements such as 'the human brain is a computer' and 'a computer can think' had a different meaning then than they do now. The design of the computer was in fact resolved *after* the idea of artificial neural networks was introduced. Thomas Haigh and Mark Priestley have also noted that the influence of the Turing machine model on the actual design of the computers is overestimated. Von Neumann was among the first to propose a viable design for the Turing machine, which has since become known as the 'von Neumann architecture'. However, according to Haigh and Mark, von Neumann mostly copied

In order to put the history of AI in perspective, it is important at this point to clarify the difference between deductive and inductive logic. Since Leibniz's idea of a *calculus ratiocinator*, the modern project of pursuing machine intelligence has been founded on the postulate that human logic can be expressed by propositional logic ('if x, then y is true/false'), which is equivalent to Boolean logic (AND, OR, and NOT operators). A proposition expressed according to this formal logic can be easily encoded into a mechanism made of rotating gears, electric relays, or electronic gates, such as valves or transistors. On a closer look, this type of logic is pursuing a linear form of rationality that replicates the linearity of written language and symbol manipulation according to the rules of *deductive inference* – an approach well exemplified by the Turing machine and the way it executes instructions from a one-dimensional tape. Symbolic AI, expert systems, and inference engines are all examples of this tendency of deductive machines that continued until the 1980s. On the other hand, artificial neural networks – along with all machine learning algorithms – are examples of inductive machines. Whereas deductive logic is the application of a rule, reasoning from the general to the particular, inductive logic involves reasoning from the particular to the general, thereby forming rules of classification. The canonical example is the movement from the discovery that 'each human being dies' to the definition of the rule that 'all human beings are mortal'. This opposition between deductive and inductive logic is key to understanding not only the Gestalt controversy but also the rise of machine learning.

According to Heims's account, in preparation for the highly anticipated visit of the Gestalt scholar Köhler to the Macy conferences in New York, Klüver issued a challenge to the gathered scientists: to formulate a theory on 'how an automaton could perceive Gestalten'. Klüver's challenge was taken up by McCulloch and Pitts, who, with the financial support of the Macy and Rockefeller foundations, were seeking to move

---

McCulloch and Pitts's 'biological language' in describing the EDVAC computer in 1945. He called the main components of the EDVAC 'organs' and its storage 'memory', and deployed 'an abstract neuron-inspired notation to represent computing circuits'. See Thomas Haigh and Mark Priestley, 'Von Neumann Thought Turing's Universal Machine Was "Simple and Neat", but That Didn't Tell Him How to Design a Computer', *Communications of the ACM* 63, no. 1 (2019): 26–32; and John von Neumann, *Theory of Self Reproducing Automata*, Urbana: University of Illinois Press, 1966, 45.

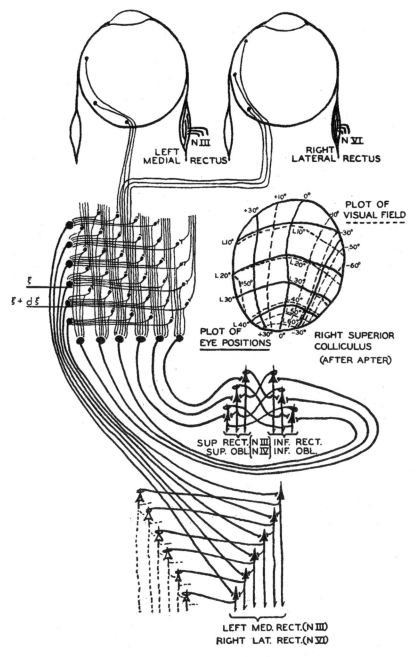

Figure 7.1. Diagram of the superior colliculus of the midbrain. Warren McCulloch and Walter Pitts, 'How We Know Universals: The Perception of Auditory and Visual Forms', *Bulletin of Mathematical Biophysics* 9, no. 3 (1947): 141.

beyond their 1943 proof that a finite network of simplified model-neurons could compute anything that can be unambiguously stated in words. Their aim, this time, was to develop neural mechanisms for specific brain functions such as perception, ideally with enough precision to be tested experimentally.[13]

As a result of the discussion, in 1947, McCulloch and Pitts published their paper 'How We Know Universals: The Perception of Auditory and Visual Forms'.[14] Compared to the previous one from 1943, here the concern is not the computation of propositions according to a linear logic but, rather, the recognition of a pattern in a bidimensional matrix. McCulloch and Pitts conceived a 'device' by which 'numerous nets, embodied in special nervous structures, serve to classify information according to useful common characters'.[15] What would this 'device' look like? Given the technical development of the late 1940s, it had to be made, for instance, of photoreceptors that could turn a visual input into a digital image – that is, a grid of pixels measured by numerical values. Without going into detail about colour, presumably such a grid would have to have been encoded in binary digits: value 1 for black pixels, value 0 for white pixels. McCulloch and Pitts's device for pattern recognition was a computing network that would resolve a grid of binary numbers into the binary output 1 (true) if a pattern was recognised or 0 (false) if a pattern was not recognised.

In mathematical terms, a device for pattern recognition is an algorithm that computes the same output for any input that a human classifies as belonging to the same type of patterns. With their mathematical device, McCulloch and Pitts wanted to challenge the Gestalt school's belief in the uniqueness of human cognition and demonstrate that perception can be described algorithmically and automated. Their original exposition of an artificial neural network for pattern recognition reads:

> Numerous nets, embodied in special nervous structures, serve to classify information according to useful common characters. In vision

---

13 Heims, *Cybernetics Group*, 224, 231.

14 Warren McCulloch and Walter Pitts, 'How We Know Universals: The Perception of Auditory and Visual Forms', *Bulletin of Mathematical Biophysics* 9, no. 3 (1947): 127–47.

15 McCulloch and Pitts, 'How We Know Universals', 127.

they detect the equivalence of apparitions related by similarity and congruence, like those of a single physical thing seen from various places. In audition, they recognize timbre and chord, regardless of pitch. The equivalent apparitions in all cases share a common figure and define a group of transformations that take the equivalents into one another but preserve the figure invariant. So, for example, the group of translations removes a square appearing at one place to other places; but the figure of a square it leaves invariant. These figures are the *geometric objects* of Cartan and Weyl, the *Gestalten* of Wertheimer and Köhler. We seek general methods for designing nervous nets which recognize figures in such a way as to produce the same output for every input belonging to the figure.[16]

According to McCulloch and Pitts, it would be possible to compute the principles of Gestalt recognition even when they seem to involve a profound power of abstraction, as in the phenomenon of closure (when, for example, to use again the above-cited example, a triangle is recognised from incomplete lines and non-connected points). Although it may appear reductionist, McCulloch and Pitts in fact proposed a complex dimorphism of dimension between perception and cognition. Gestalt theorists maintained that perception and cognition exist along a continuous isomorphism: the spatial relations of an object are isomorphic with its percept, and thus with its mental image. In contrast, McCulloch and Pitts suggested the possibility of transforming the two-dimensional relations of an image into a one-dimensional representation – basically a code, or logical proposition – without undermining the content or quality of information.

This is a crucial passage in the history of connectionism and AI: McCulloch and Pitts argued that the isomorphism between an image and its logical representation is not necessary. The form of a triangle, for instance, does not need to be memorised as an isomorphic triangle in some parts of the brain but can be distributed across different dimensions (a topological transformation also termed 'internal representation' in deep learning).[17] The two scientists advanced this hypothesis against

---

16 Ibid.
17 It is also for this reason that Kölher accused McCulloch and Pitts's neural model of old-school 'dualism'. Jeffress, *Hixon Symposium*, 69.

the Gestalt school to argue that a visual manifold can be computed by a linear information machine such as that of Turing:

> This point is especially to be taken against the Gestalt psychologists, who will not conceive a figure being known save by depicting it topographically on neuronal mosaics, and against the neurologists of the school of Hughlings Jackson, who must have it fed to some specialized neuron whose business is, say, the reading of squares. That language in which information is communicated to the homunculus who sits always beyond any incomplete analysis of sensory mechanisms and before any analysis of motor ones neither needs to be nor is apt to be built on the plan of those languages men use toward one another.[18]

McCulloch and Pitts's paradigm of perception and cognition was as much as atomistic as multidimensional: it was based on the idea that in order to 'perceive' and 'understand', a biological or artificial neural network can dismember a manifold of two dimensions and project it onto a different configuration of dimensions than the original ones. In a visionary way, they challenged the famous dictum by Gestalt theory: 'The whole is other than the sum of its parts.'[19] They sought to replace Gestalt theory's holistic complexity with another kind of multidimensionality which was purely mathematical.

What the frog's eye tells the frog's brain

At the Hixon symposium, Köhler was the lone delegate representing the Gestalt school. He participated not to defend a mystic power of the mind but to present physical measurements of brain activity during visual perception, precisely 'records of electric currents which have been taken from the heads of human subjects under conditions of pattern vision.'[20] Köhler's study demonstrated that Gestalt theorists were not

---

18 McCulloch and Pitts, 'How We Know Universals'.
19 Originally formulated by Kurt Koffka. Köhler, *Principles of Gestalt Psychology*, New York: Harcourt & Brace, 1935, 176.
20 Wolfgang Köhler, 'Relational Determination in Perception', in Jeffress, *Hixon Symposium*, 200.

opposed to the quest for neural correlates of perception; they simply pursued a traditional method of scientific testing rather than the thinking by analogy of cyberneticians. As Heims has also noted:

> Gestalt psychologists did not capitulate to mysticism, political reaction and vitalism, all of which were connected with the hunger for wholeness. They insisted on empirical studies of phenomena and sought fundamental laws describing structural characteristics of experience, even as they supported the popular opposition to atomistic and mechanistic analyses of the world.[21]

Specifically, Köhler relied on the model of force fields, which was derived from physics and inspired, among others, by his close friend Max Planck, to explain the phenomenon of visual perception happening between eye and brain.

At the Hixon symposium, the *physical hypothesis* of force fields underlying Gestalt perception clashed with the *computational analogy* of brains as finite-state automata. Köhler attacked McCulloch's reductionist approach to the brain, arguing that nerve impulses, when seen and measured by a histologist or neurophysiologist, do not look like logical propositions but simply like ... impulses![22] Against cybernetics' view of the brain as 'machine arrangements', Köhler proposed the theory of force fields to explain a structural continuity of form (isomorphism) between the stimulus of perception, its neural correlates, and the higher faculties of cognition (such as the faculty of insight):[23]

> Continuity is a *structural* trait of the visual field. It is also a *structural* fact that in this field circumscribed particular percepts are segregated as patches, figures and things. In both characteristics, we have found, the macroscopic aspect of cortical processes resembles visual

---

21 Heims, *Cybernetics Group*, 108.
22 See Köhler's comment to McCulloch in Jeffress, *Hixon Symposium*, 66.
23 Wolfgang Köhler, 'Review of *Cybernetics or Control and Communication in the Animal and the Machine* by Norbert Wiener', *Social Research* 18, no. 1 (1951): 127. On isomorphism, see Abraham Luchins and Edith Luchins, 'Isomorphism in Gestalt Theory: Comparison of Wertheimer's and Köhler's concepts', *Gestalt Theory* 21, no. 3 (1999): 208–34.

experience. To this extent, therefore, vision and its cortical correlate are isomorphic.[24]

In contrast with the gung-ho enthusiasm of cyberneticians for neural correlates resembling logical operators, Köhler was cautious about his own findings.[25] Moreover, contrary to what the term may suggest, Gestalt 'isomorphism' was not a monotonic notion but one that included principles of plasticity such as self-organisation and self-repair in the broader sense.[26] The Gestalt school considered the idea of the mind as a finite-state automaton unfit for describing these higher capacities of the mind for synthesis and abstraction through self-organisation.

On the other side of the camp, the most vocal and virulent anti-Gestaltist was Norbert Wiener, who called holism a 'pseudo-scientific bogy'. Accordingly, Wiener argued that 'if a phenomena can only be grasped as a whole and is completely unresponsive to analysis, there is no suitable material for any scientific description of it; for the whole is never at our disposal'.[27] Wiener shared McCulloch and Pitts's computational view of the mind ideologically, with no hesitation:

It is a noteworthy fact that the human mind and animal nervous systems, which are known to be capable of the work of a computation

---

24 Wolfgang Köhler, *The Place of Value in a World of Facts*, New York: Liveright, 1938, 217. See Wolfgang Köhler, *Dynamics in Psychology*, New York: Liveright, 1940.

25 'If it were really demonstrated that pattern vision is a matter of continuity physics, then we should be seriously tempted to believe that this holds also for analogous facts in other parts of psychology. Just because I realize that we are dealing with a major problem of psychophysics in general, I feel inclined to move slowly. It may be that the cortical correlate of pattern vision has now become accessible to experimentation. But we can not yet be sure that it has.' See Köhler, 'Relational Determination in Perception', 229.

26 On organicism and plasticity see Kurt Goldstein, *Der Aufbau des Organismus: Einführung in die Biologie unter besonderer Berücksichtigung der Erfahrungen am kranken Menschen*, Den Haag: Nijhoff, 1934, translated as *The Organism: A Holistic Approach to Biology Derived from Pathological Data in Man*, New York: American Book Company, 1939.

27 Norbert Wiener, 'Some Maxims for Biologists and Psychologists' (1950), in *Collected Works with Commentaries*, vol. 4, *Cybernetics, Science, and Society; Ethics, Aesthetics, and Literary Criticism; Book Reviews and Obituaries*, ed. P. Masini, Cambridge, MA: MIT Press, 1985, 454–6.

system, contain elements which are ideally suited to act as relays. These elements are the so-called *neurons* or nerve cells.[28]

The idea of artificial neurons and the project of automating Gestalt perception had already been discussed when, in 1948, Wiener published *Cybernetics*, which provided the first summa of digital computation and information theories. In 1951, a few years after the Hixon symposium, Köhler wrote a harsh review of Wiener's book and its chapter about 'Gestalt and Universals', in which he argued that while it is true that machines are better than humans at calculation, such calculation does not constitute thinking or, indeed, capacity of insight (*Einsicht*) into a problem:

> In the relation of human beings to the computing machines, thinking in the proper sense of the term appears to remain the task of the former. Excellent engineers and mathematicians build into such instrument mechanically forced ways of operation which may serve as factually reliable substitutes for certain quantitative activities of the human mind such as addition and multiplication. It is not astonishing that, so far as speed is concerned, the substituted operations of the machines are far superior to anything that human brains can achieve. At the same time, these operations appear to be generically different from those of a human being who is occupied with a mathematical problem . . . The machines do not know, because among their functions there is none that can be compared with insight into the meaning of a problem.[29]

In the mid-1950s, neurophysiologist Jerome Lettvin and biologist Humberto Maturana invited McCulloch and Pitts to contribute to their studies on the neurons of the frog's eye. Their conclusion would be published in 1959 in the famous paper 'What the Frog's Eye Tells the Frog's Brain', which for many represents the final nail in the coffin of the Gestalt controversy.[30]

---

28 Norbert Wiener, *Cybernetics: Or Control and Communication in the Animal and the Machine*, Cambridge, MA: MIT Press, 1961 [1948], 120.

29 Köhler, 'Review of *Cybernetics*', 128.

30 J. Y. Lettvin et al., 'What the Frog's Eye Tells the Frog's Brain', *Proceedings of the IRE* 47, no. 11 (1959): 1940–51.

Their research moved from the simple observation that the optical nerve of higher animals is composed of fewer fibres than the retina. This suggests that the stimulus must undergo compression before it reaches the brain. The question that arises is what exactly happens to the stimulus between the retina, the optical nerve, and the visual cortex? The scientists took the frog as their model organism. They measured the stimuli in the optical nerve using an electrode, while presenting high-contrast shapes to the animal's eye. In this way, they recorded four types of neural behaviours, or 'operations': '1) sustained contrast detection; 2) net convexity detection; 3) moving edge detection; 4) net dimming detection'.[31] By way of illustration, let us consider the second operation – net convexity detection – which is perhaps the most intuitive to understand: it is about the perception of a small object roaming across the visual field, like a flying insect, which is crucial for food retrieval and the survival of the frog. As such, it was colloquially called the 'bug detector':

Against the Gestalt theory's primacy of brain functions, the authors argued that the eye already performs basic tasks of cognition, such as pattern recognition, and sends signals to the brain that are already well-formed *concepts* and not just *percepts*. They contended that the eye is already employing the language of 'complex abstractions' rather than being simply a medium of sensations. They concluded that 'the eye speaks to the brain in a language already highly organized and interpreted, instead of transmitting some more or less accurate copy of the distribution of light on the receptors'.[32] What the eye sends to the brain would not be of a form analogous to the percept (which would maintain *Gestalt* proportions) but rather a semantic proposition – a piece of information interpreting the pattern present in the percept. The collective article also integrated findings from McCulloch and Pitts's 1947 paper, including the observation that an image can be logically defined and 'perceived' by the calculation of the relations between its elements according to a spatial logic.

> By transforming the image from a space of simple discrete points to a congruent space where each equivalent point is described by the intersection of particular qualities in its neighborhood, we can then give the

---

31 Ibid., 1943.
32 Ibid., 1950.

image in terms of distributions of combinations of those qualities. In short, every point is seen in definite contexts. The character of these contexts, genetically built in, is the physiological synthetic a priori.[33]

A digital image is comprised not simply of discrete points but of spatial relations that can be described as 'distributions of combinations'. Notably, the authors seem to suggest here that it is not the neural network per se which comprises the 'physiological synthetic a priori' of thought but rather the spatial logic that the neural network computes.

The biomorphism of the early artificial neural networks (in this case, the imitation of biological neural networks) must eventually be questioned. The human organ that McCulloch and Pitts most often investigated and envisioned as a network of logical operators was not the brain but the eye; and, of the eye, their artificial neural networks implicitly inherited a specific hierarchy of behaviour. In McCulloch and Pitts's 1947 essay, the cooperation among the nodes of the artificial neural network was supposed to resolve the combinatorial geometry of the field of vision, not the propositional logic of a syllogism, as in their 1943 publication. In the Gestalt controversy, then, one does not find a model of cognition to be mechanised but rather several models of perception, since the elements of the perceptual field, rather than mental states, were represented in computational terms. The development of artificial neural networks continued by mostly studying the neurons of the eye rather than the brain: from Lettvin's frogs to David Hubel and Torsten Wiesel's cats (which inspired convolutional neural networks such as AlexNet), the history of connectionist AI is indebted to experiments on animals and their organs of vision.[34]

From a mathematical point of view, the Gestalt controversy was, on one hand, about the transformation of an image into a logical construct and, on the other, about the fact that a logical construct of human reasoning may share the probabilistic features of image perception.[35]

---

33 Ibid.

34 D. H. Hubel and T. N. Wiesel, 'Receptive Fields of Single Neurones in the Cat's Striate Cortex', *Journal of Physiology* 124, no. 3 (1959): 574–91.

35 Historian of science Peter Galison has investigated the relation between the 'mimetic' and 'logical' traditions in the construction of scientific images. See Peter Galison, *Image and Logic: A Material Culture of Microphysics*, Chicago: University of Chicago Press, 1997.

Reading between the lines, the Gestalt controversy was already pointing towards a logical form that would only emerge properly in twenty-first-century machine learning: the idea of a spatial and statistical dimension of information beyond the linear modality of the Turing machine. Logic gates and perceptual fields – the basic concepts of cybernetics and the Gestalt school respectively – were two polarised ideals that the evolution of AI eventually overcame and synthesised into more and more abstract constructs: today the inner logic of machine learning is described, for instance, through entities such as multidimensional vectors, latent spaces, and statistical distributions. The visual matrix (and, with it, the picture plane of modern iconology) has since then evolved into these advanced logical forms, into vast ramifications of statistical inference yet to be fully understood.

The language of the brain is not the language of mathematics

As the epistemologist David Bates has noted, the Gestalt school's scepticism about computational reductionism was nevertheless received by a central protagonist of US cybernetics: the polymath John von Neumann. A key figure of the Manhattan Project and fervent anti-Communist,[36] von Neumann occupied a role of mediator in the controversy. In so doing, he effected a synthesis of Gestalt theory and logical neural networks and, in this way, opened the terrain for Rosenblatt's idea of statistical neural networks. He maintained a strong interest in the mathematical analysis of biological neural networks without claiming either that the brain is a machine or that a machine can perfectly simulate a brain. Instead, he argued that, considering the number of their units (neurons in brains relative to switches in machines), the scale of human cognition cannot be compared with mechanical computation: 'The number of nerve cells in the human central nervous system has been estimated to be 10 billion . . . and nobody has seen a switching organ of 10 billion units; therefore one would be comparing two unknown

---

36 David Bates, 'Creating Insight: Gestalt Theory and the Early Computer', in *Genesis Redux: Essays in the History and Theory of Artificial Life*, ed. Jessica Riskin, Chicago: University of Chicago Press, 2007, 237–59.

objects.'³⁷ As we will see, ultimately von Neumann did not reduce the brain to the computer – the biological to the logical – but instead envisioned in-between models and implementations to translate one domain into the other.

Von Neumann agreed with McCulloch and Pitts that 'anything that you can describe in words can also be done with the neuron method' but disagreed 'that any circuit you are designing in this manner really occurs in nature'.³⁸ While he sympathised with McCulloch and Pitts's effort to axiomatise biological neurons into formal ones, he also agreed with the Gestalt school that the systemic nature of form perception is such that it cannot be easily described in propositional terms and deductive logic, unless by exploding the size of descriptions into lengthy and verbose code scripts. As von Neumann explains, arithmetic logic carries manifest limits of scale in its representation of the world:

> Suppose you want to describe the fact that when you look at a triangle you realize that it's a triangle, and you realize this whether it's small or large. It's relatively simple to describe geometrically what is meant: a triangle is a group of three lines arranged in a certain manner. Well, that's fine, except that you also recognize as a triangle something whose sides are curved, and a situation where only the vertices are indicated, and something where the interior is shaded and the exterior is not. You can recognize as a triangle many different things, all of which have some indication of a triangle in them, but the more details you try to put in a description of it the longer the description becomes. [With] respect to the whole visual machinery of interpreting a picture, of putting something into a picture, we get into domains which you certainly cannot describe in those [logical] terms.³⁹

Contra McCulloch and Pitts, von Neumann contended that the brain can recognise the complexity of any image precisely because of its probabilistic logic. Arithmetic logic would be then a simplification and approximation of the brain's actual workings and could only partially

---

37 Von Neumann, *Theory of Self Reproducing Automata*, 37. See also John von Neumann, 'The General and Logical Theory of Automata', in Jeffress, *Hixon Symposium*, 3.
38 Von Neumann, *Theory of Self Reproducing Automata*, 46.
39 Ibid., 46–7.

grasp its features. As a result, von Neumann defined arithmetic logic as a metalanguage ('secondary language' or 'short code') that is efficient but not effective at fully describing the underlying probabilistic language of the brain ('primary language' or 'complete code'). He also observed, similarly to Köhler, that the nervous system speaks this primary language using the statistical properties of stimuli ('complete code'), rather than exact markers ('short code').

Given such probabilistic dynamics, von Neumann concluded, counterintuitively, that in the nervous system a 'deterioration' of arithmetic precision can actually result in 'an improvement in logics'.[40] He highlighted, among others, that biological neural networks operate with an error tolerance that would set any deterministic computing machine out of joint. A single mistake in a computer programme and the whole machine halts. However, such mistakes never trouble organisms. It is, then, the same probabilistic nature of organisms that make them fault redundant and able to perceive fuzzy and complex figures effortlessly:

> The fact that natural organisms have such a radically different attitude about errors and behave so differently when an error occurs is probably connected with some other traits of natural organisms, which are entirely absent from our automata. The ability of a natural organism to survive in spite of a high incidence of error (which our artificial automata are incapable of) probably requires a very high flexibility and ability of the automaton to watch itself and reorganize itself. And this probably requires a very considerable autonomy of parts. There is a high autonomy of parts in the human nervous system. This autonomy of parts of a system has an effect which is observable in the human nervous system but not in artificial automata.[41]

There is today consensus that biological neural networks operate in spite of errors, faults, and damages – and precisely also thanks to them.[42] Cyberneticians encountered this idea of neuroplasticity in the cognitive sciences of the time, in particular in the work of the neurologists Karl

---

40 John von Neumann, *The Computer and the Brain*, Silliman Memorial Lectures, New Haven, CT: Yale University Press, 1958, 80.

41 Von Neumann, *Theory of Self Reproducing Automata*, 73.

42 For the influence of this idea on the genesis of deep learning, see Yann LeCun et al., 'Optimal Brain Damage', in *Neural Information Processing Systems* 2 (1989): 598–605.

Lashley (who also presented at the Hixon symposium) and Kurt Goldstein (who had emigrated to the US as part of the Gestalt school diaspora). Lashley tested the effect of artificially induced brain injuries on the memory of laboratory rats, while Goldstein studied the capacity of World War I veterans to reorganise their memories after real brain traumas.[43] However, it was Goldstein rather than Lashley who provided a more systematic model of neuroplasticity. His theory of the brain's capacity to reorganise itself after a trauma was also closely related to the debate on memory localisation. A thesis key to a pioneer of holistic neurology, Constantin von Monakow, was that memory is distributed rather than localised, and this is why it can be recovered after a brain injury (see chapter 8). Von Neumann echoed such theories:

> The main difficulty with the memory organ is that it appears to be nowhere in particular. It is never very simple to locate anything in the brain, because the brain has an enormous ability to re-organize. Even when you have localized a function in a particular part of it, if you remove that part, you may discover that the brain has reorganized itself, reassigned its responsibilities, and the function is again being performed.[44]

It was this idea of a distributed memory in human brains that suggested the delocalisation of memory in machines. Neural networks, with their distributed architecture, were the perfect candidates to accomplish this. As Rosenblatt acknowledged in *Neurodynamics*, von Neumann's comment on distributed memory was a main inspiration for the perceptron (see chapter 9).[45]

---

43 The chapter on the 'pathologies of machines' in Wiener's *Cybernetics* is an echo of these debates on neuroplasticity. See also Matteo Pasquinelli (ed.), *Alleys of Your Mind: Augmented Intelligence Trauma*, Lüneburg: Meson Press, Leuphana University, 2015, introduction; David Bates, 'Unity, Plasticity, Catastrophe: Order and Pathology in the Cybernetic Era', in *Catastrophe: History and Theory of an Operative Concept*, ed. Andreas Killen and Nitzan Lebovic, Berlin: De Gruyter, 2014, 252; Matteo Pasquinelli, 'What an Apparatus Is Not: On the Archeology of the Norm in Foucault, Canguilhem, and Goldstein', *Parrhesia* 22 (May 2015): 79–89.

44 Von Neumann, *Theory of Self Reproducing Automata*, 49. See also *The Computer and the Brain*, 63–8.

45 Frank Rosenblatt, *Principles of Neurodynamics: Perceptrons and the Theory of Brain Mechanisms*, report, Buffalo, NY: Cornell Aeronautical Laboratory, 1961, 10. On Lashley, see page 4.

Seen in perspective, von Neumann pursued a different method of inquiry compared to the other cyberneticians. Against the Platonism and intuitionism then popular also in engineering, von Neumann maintained a constructivist perspective on language, logic, and mathematics. He believed that these concepts were not inherent or innate, but rather products of historical development.

> It is only proper to realize that language is largely a historical accident. The basic human languages are traditionally transmitted to us in various forms, but their very multiplicity proves that there is nothing absolute and necessary about them. Just as languages like Greek or Sanskrit are historical facts and not absolute logical necessities, it is only reasonable to assume that logics and mathematics are similarly historical, accidental forms of expression. They may have essential variants, i.e. they may exist in other forms than the ones to which we are accustomed. Indeed, the nature of the central nervous system and of the message systems that it transmits indicate positively that this is so.[46]

This historical relativisation of mathematics and logic is crucial for understanding von Neumann's approach to the computational theory of the mind and the project of building thinking automata. Von Neumann held a more sophisticated position than functionalism, or the thesis of the *multiple realisability of the mind*, which maintains that the same operative model of intelligence can be implemented across different substrates made of neurons, relays, transistors, and so forth. Instead, von Neumann promoted a method of mutual implementation between artificial and natural worlds, that is, a method of *modelling*.[47] Of course, von Neumann was positive about using computers as model machines for studying the brain.

Even so, at the end of his life, when he was invited by Yale University to deliver the Silliman Lectures in 1956, he sought to clarify this point without ambiguity. His lecture series 'The Computer and the Brain' concludes with the declarative title 'The Language of the Brain Not the

---

46 Von Neumann, *The Computer and the Brain*, 82.
47 John von Neumann, 'The General and Logical Theory of Automata', in Jeffress, *Hixon Symposium*, 3.

Language of Mathematics'. In this lecture series, he recognised a key role to the 'secondary language' of mathematics in the knowledge of the 'primary language' of the brain. But, rather than reiterating the computationalism of McCulloch, Pitts, and Wiener, von Neumann – himself no romantic – made at the end a remarkable intervention: he reversed the relation between logic and nature, computer and brain, to the point of suggesting that the study of neurophysiology could one day reshape logic altogether. In the introduction to the lectures's anthology, shortly before his death in 1957, presciently, von Neumann wrote: 'I suspect that a deeper mathematical study of the nervous system [in fact] may alter the way in which we look on mathematics and logics proper.'[48]

Von Neumann's final thoughts were a recognition of the dialectical relation of any system of thought with nature and the external world. They came to terms with the bounds of any system of formalisation and acknowledged the material constraints, including historical ones, of technological and scientific abstractions. These insights surpassed the standard reductionism of the cybernetic mentality. Models of minds and machines do not need to match each other; they often operate through modelling the world rather than reducing it to fixed representations. An epistemology which would acknowledge the flexibility and limitations of modelling is recommended, and yet not a sufficient condition for a progressive philosophy of the mind. As we will see in the next chapter, in fact, the neoliberal economist Friedrich Hayek attempted to turn model thinking into a core principle of economic individualism.

---

48 Von Neumann, *The Computer and the Brain*, 2.

8
Hayek and the Epistemology of Connectionism

Mind thus becomes to me a continuous stream of impulses, the significance of each and every contribution of which is determined by the place in the pattern of channels through which they flow within the pattern of all available channels – with newly arriving afferent impulses, set up by external or internal stimuli, merely diverting this flow into whatever direction the whole flow is disposed to move . . . I liked to compare this flow of 'representative' neural impulses, largely reflecting the structure of the world in which the central nervous system lives, to a stock of capital being nourished by inputs and giving a continuous stream of outputs – only fortunately, the stock of this capital cannot be used up.
    Friedrich Hayek, 'The Sensory Order after 25 Years' [1977][1]

Homo sapiens is about pattern recognition . . . Both a gift and a trap.
    William Gibson, *Pattern Recognition*, 2003[2]

---

1 Friedrich Hayek, 'The Sensory Order after 25 Years', in *Cognition and the Symbolic Processes*, vol. 2, ed. W. B. Weimer and D. S. Palermo, Hillsdale, NJ: Erlbaum, 1982, 287–93. Originally a lecture delivered in 1977.

2 William Gibson, *Pattern Recognition*, New York: Putnam's Sons, 2003, 22.

Introducing the classifier

It was not a cybernetician but a neoliberal economist who provided the most systematic treatise on connectionism, or, as it would be later known, the paradigm of artificial neural networks.[3] In his 1952 book *The Sensory Order*, Friedrich Hayek propounded a connectionist theory of the mind already far more advanced than the theory of symbolic AI, whose birth is, redundantly, celebrated anno 1956 with the exalted Dartmouth workshop.[4] In *The Sensory Order*, Hayek provided a synthesis of Gestalt principles and Warren McCulloch and Walter Pitts's idea of neural networks to describe 'the nervous system as an instrument of classification'.[5] He went so far as to speculate about the possibility of a device fulfilling a similar function, describing (in the jargon of today's machine learning) a classifier algorithm. In 1958, Frank Rosenblatt defined the perceptron (the first operative artificial neural network for pattern recognition) as 'connectionist' and acknowledged that the work of 'Hebb and Hayek' was 'the most suggestive' for his own.[6] While Donald Hebb was a neuropsychologist famous for the theory of brain cell assemblies – a doctrine of neuroplasticity that is encapsulated in the dictum 'Neurons that fire together, wire together' seen in chapter 6 – Hayek was an economist who studied the self-organisation of the mind in a similar way but in order to support a political belief: namely, the spontaneous order of markets. The perception that Hayek invented connectionism, however, is a simplification that overlooks his debt to the neurology and cybernetics of the time. One might better say that Hayek stole pattern recognition and transformed it into a neoliberal principle of market regulation.

---

3 Donald Hebb introduced the term 'connectionist' in his 1949 book *The Organization of Behavior: A Neuropsychological Theory*, New York: Wiley & Sons, 1949. Rosenblatt adopted the term in 1958 to define his theory of artificial neural networks: Frank Rosenblatt, 'The Perceptron: A Probabilistic Model for Information Storage and Organization in the Brain', *Psychological Review* 65, no. 6 (1958).

4 Friedrich Hayek, *The Sensory Order: An Inquiry into the Foundations of Theoretical Psychology*, Chicago: University of Chicago Press, 1952. *The Sensory Order* is a development of the manuscript 'Beiträge zur Theorie der Entwicklung des Bewusstseins' ('Contributions to the Theory of the Development of Consciousness') which Hayek wrote in German as early as 1920.

5 Hayek, *The Sensory Order*, chapter 3, 55.

6 Rosenblatt, 'The Perceptron'.

Hayek began work on his theory of the mind in 1920, when he was an assistant in the laboratory of neuropathologist Constantin Monakow in Zurich, and continued developing it across a long list of publications throughout his career.[7] He provided an impressive synthesis of ideas, from neurophysiology, holistic neurology, Gestalt psychology, system theory, empirio-criticism, and cybernetics – although he mobilised this armamentarium of cognitive science with the purpose of making neoliberal principles look natural and universal.[8] A striking example of this is that Hayek described the decentralisation of knowledge across the market in the same way in which Constantin von Monakow and Kurt Goldstein's theories of neuroplasticity described the decentralisation of cognitive functions across the brain.

As seen in chapter 6, between the 1940s and the 1960s, the theory of self-organisation in markets contributed to the theories of self-organisation in computing networks, and vice versa. It must be said, however, that Hayek's theory of the market's spontaneous order was part of an ideological coup d'état. Indeed, nothing looked less spontaneous than a market order within the sphere of influence of a nuclear superpower.[9] As noted earlier, historians of science and technology usually stress the influence of US military funding on the development of cybernetics and artificial intelligence. However, another front of the Cold War has to be acknowledged to complete the picture: the formation of neoliberal doctrines in response to the socialist calculation debate (described below) and Keynesian

---

7 See Friedrich Hayek, 'Scientism and the Study of Society' part 1, *Economica* 9, no. 35 (1942): 267–91; Friedrich Hayek, 'The Theory of Complex Phenomena', in *The Critical Approach to Science and Philosophy: Essays in Honor of Karl R. Popper*, ed. Mario A. Bunge, New York: The Free Press of Glencoe, 1964; Friedrich Hayek, 'Rules, Perception and Intelligibility', *Proceedings of the British Academy* 48 (1963): 321–44; Friedrich Hayek, 'The Primacy of the Abstract', in *Beyond Reductionism: The Alpbach Symposium*, ed. Arthur Koestler and J. R. Smythies, London: Hutchinson, 1969.

8 Discussing Hayek's legacy in the conceptualisation of information, Philip Mirowski and Edward Nik-Khah have noticed that 'the place of information in economics was broached in heated disputes over the politics and possibilities of socialism'. Philip Mirowski and Edward Nik-Khah, *The Knowledge We Have Lost in Information: The History of Information in Modern Economics*, Oxford: Oxford University Press, 2017, 65.

9 See Hayek's 1977 visit to Chile and the meeting with the dictator Augusto Pinochet.

policies.¹⁰ Just as much as the decentralised topology of the Arpanet military network (the precursor of the internet) was designed as a reaction to Soviet military threat, Hayek's connectionism was conceived, among other stimuli, as a response to socialist centralised planning and Keynesianism.¹¹ Reading Hayek through this lens helps to illuminate the influence of economic rationality on the early paradigms of artificial intelligence and trace the circulation of those ideas through models of minds, markets, and machines in the post–World War II years, but also to register the influence of political and social forces in the making of such models. It was a competitive market network that gave form to Hayek's neural networks, which were elevated to techniques for price calculation because, as Hayek confessed in the epigraph to this chapter, they were implicitly envisioned as 'a stock of capital being nourished by inputs and giving a continuous stream of outputs'.¹² In this sense, Hayek's theory of the mind was but a variant of *mercantile connectionism*.

This chapter aims to put Hayek's epistemological project 'on its feet', so to speak, showing how his connectionist theory of the mind was used to shore up a specific (ideological) view of the market. This will require a schematic reconstruction of Hayek's argument from his economic paradigm backwards to his theory of cognition. Hayek tried to forward the following lines of argumentation: (1) that the economic problem is about the limited knowledge of free individuals which establish the optimal price of commodities on the basis of incomplete information; (2) that knowledge is acquired through the act of classification, or pattern recognition – that is, the universal faculty to make categories out of perceptions that appear different and incomplete; (3) that classification happens via the self-organisation of connections in the brain, or neural networks – in other words, that knowledge is not made of propositions and representations but is performed by a topology of connections to take decisions (to classify something within a class or not); and (4) that the mind is a dynamic mental order of

---

10 On Cold War rationality, see Paul Erickson et al., *How Reason Almost Lost Its Mind: The Strange Career of Cold War Rationality*, Chicago: University of Chicago Press, 2013.

11 Slava Gerovitch, 'InterNyet: Why the Soviet Union Did Not Build a Nationwide Computer Network', *History and Technology* 24, no. 4 (December 2008): 335–50.

12 Hayek, 'The Sensory Order after 25 Years'.

connections that is related but not identical to the external order – under which logic, knowledge is not a rigid representation but an approximate model of the world constantly rearranging itself. Eventually, in Hayek's political intention, connectionism and neural networks provide a relativist paradigm for justifying the 'methodological individualism' of neoliberalism.[13]

The decentralised and tacit rationality of the market

In 1945, Hayek intervened in the famous socialist calculation debate with the essay 'The Use of Knowledge in Society'. Ludwig von Mises of the Austrian school of economics had initiated the debate, arguing that in the absence of commodity prices as a unit of account, rational economic calculations would be impossible under the centralised bureaucracy of socialist economies. On the other side of the debate, it happened that Marxist economists such as Oskar Lange were questioning the importance of units of calculation such as money and labour time in the formation of prices. Hayek agreed with his mentor Mises but framed the anti-socialist argument differently: the economic order was, he claimed, an issue of spontaneous knowledge rather than of mathematical exactitude. Hayek saw the pricing of commodities as a spontaneous order emerging from tacit knowledge – that is, as 'a problem of the utilization of knowledge which is not given to anyone in its totality'. For this reason, neither centralised institutions nor technical apparatuses of calculation could grasp and embody such knowledge efficiently. Hayek's famous passage on the decentralised rationality of the market reads:

> The peculiar character of the problem of a rational economic order is determined precisely by the fact that the knowledge of the circumstances of which we must make use never exists in concentrated or integrated form but solely as the dispersed bits of incomplete and

---

13 See Francesco Di Iorio, *Cognitive Autonomy and Methodological Individualism: The Interpretative Foundations of Social Life*, Berlin: Springer, 2015; Francesco Di Iorio, 'The Sensory Order and the Neurophysiological Basics of Methodological Individualism', in *The Social Science of Hayek's the Sensory Order: Advances in Austrian Economics*, vol. 13, ed. William N. Butos, London: Emerald, 2010.

frequently contradictory knowledge which all the separate individuals possess. The economic problem of society is thus not merely a problem of how to allocate 'given' resources – if 'given' is taken to mean given to a single mind which deliberately solves the problem set by these 'data'. It is rather a problem of how to secure the best use of resources known to any of the members of society, for ends whose relative importance only these individuals know. Or, to put it briefly, it is a problem of the utilization of knowledge which is not given to anyone in its totality.[14]

Philip Mirowski and Edward Nik-Khah believe themselves to have found here 'the First Commandment of neoliberalism. Markets don't exist to allocate given physical resources, so much as they serve to integrate and disseminate something called knowledge.'[15] Curiously, the idea that knowledge is distributed across a system and not possessed by any single component in its totality is not an original one from Hayek but is derived from the non-localisation theory of brain functions of Monakow, with whom, as mentioned above, Hayek worked as assistant in 1920. Monakow advanced the hypothesis that cognitive functions (including memory) are not delimited in one specific part but are distributed across the whole brain. He coined the term 'diaschisis' (from the Greek for 'shocked throughout') to describe how an injured brain can recover cognitive functions through neural reorganisation.[16] Monakow's holistic model of the brain (what nowadays would be called a model of 'neuroplasticity') was further systematised by another author Hayek read and often quoted, the Gestalt neurologist Kurt Goldstein.[17]

---

14 Friedrich Hayek, 'The Use of Knowledge in Society', *American Economic Review* 35, no. 4 (1945): 519–30.

15 Mirowski and Nik-Khah, *The Knowledge*, 63.

16 Walther Riese and Ebbe C. Hoff, 'A History of the Doctrine of Cerebral Localization: Sources, Anticipations, and Basic Reasoning', *Journal of the History of Medicine and Allied Sciences* 5, no. 1 (1950): 50–71.

17 Kurt Goldstein, *The Organism: A Holistic Approach to Biology Derived from Pathological Data in Man*, New York: American Book Company, 1939. On Goldstein's influence on cybernetics, see David Bates, 'Creating Insight: Gestalt Theory and the Early Computer', in *Genesis Redux: Essays in the History and Philosophy of Artificial Life* (2007): 237–59. On Goldstein's influence on French philosophy, see Matteo Pasquinelli, 'What an Apparatus Is Not: On the Archeology of the Norm in Foucault, Canguilhem, and Goldstein', *Parrhesia* 22 (May 2015).

Hayek's idea that the market is a place of distributed knowledge therefore did not proceed from the study of economic phenomena but was first extrapolated from holistic neurology and early theories of neuroplasticity. In *The Sensory Order*, Hayek also referred to neurophysiologist Karl Lashley's idea of the brain's equipotentiality, which bears similarities to Monakow and Goldstein's:

> Certain mental processes which are normally based on impulses proceeding in certain fibres may, after these fibres have been destroyed, be relearned by the use of some other fibres. Certain associations may be effectively brought about through several alternative bundles of connexions, so that, if any one of these paths is severed, the remaining ones will still be able to bring about the result. Such effects have been observed and described under the names of 'vicarious functioning' and 'equipotentiality'.[18]

As von Neumann, among others, has suggested, holistic neurology influenced not only Hayek's idea of distributed knowledge across the market but also the architecture of distributed memory in computing machines.[19] In his 1961 book *Neurodynamics*, Rosenblatt also acknowledged Lashley's and von Neumann's remarks on the distributed architecture of the brain as one of the main inspirations for the perceptron neural network.[20]

Alongside the decentralisation of knowledge in his economic paradigm, Hayek performed another important operation of decentring: the

---

18 Hayek, *The Sensory Order*, 148.

19 'The main difficulty with the memory organ is that it appears to be nowhere in particular. It is never very simple to locate anything in the brain, because the brain has an enormous ability to re-organize. Even when you have localized a function in a particular part of it, if you remove that part, you may discover that the brain has reorganized itself, reassigned its responsibilities, and the function is again being performed. The flexibility of the brain is very great, and this makes localization difficult. I suspect that the memory function is less localized than anything else.' Von Neumann, *Theory of Self Reproducing Automata*, 49. See also von Neumann, *The Computer and the Brain*, 63–8.

20 Frank Rosenblatt, *Principles of Neurodynamics: Perceptrons and the Theory of Brain Mechanisms*, report, Buffalo, NY: Cornell Aeronautical Laboratory, 1961, 10. On Lashley: 4. See Karl Lashley, 'The Relation between Mass Learning and Retention', *Journal of Comparative Neurology* 41, no. 1 (1926): 1–58; Karl Lashley, *Brain Mechanisms and Intelligence*, Chicago: University of Chicago Press, 1929.

mobilisation of tacit knowledge.[21] Hayek took great inspiration from Gilbert Ryle's 1945 paper, 'Knowing How and Knowing That', which famously defended the status of know-how and skills against the alleged 'higher' forms of conscious and procedural knowledge.[22] Hayek writes:

> The 'know how' consists in the capacity to act according to rules which we may be able to discover but which we need not be able to state in order to obey them ... Rules which we cannot state thus do not govern only our actions. They also govern our perceptions, and particularly our perceptions of other people's actions. The child who speaks grammatically without knowing the rules of grammar not only understands all the shades of meaning expressed by others through following the rules of grammar, but may also be able to correct a grammatical mistake in the speech of others.[23]

What we recognise as purposive conduct is conduct following a rule with which we are acquainted but which we need not explicitly know. Similarly, that an approach of another person is friendly or hostile, that he is playing a game or willing to sell us some commodity or intends to make love, we recognise without knowing what we recognise it from.[24]

The holistic neurology of the time shared a similar position. For Goldstein, for instance, the unconscious is the locus not of primordial instincts that drive the conscious mind, as was the case with its Freudian predecessor, but, rather, of abstract behaviours as important as the conscious ones. By this account, the unconscious is a space of rules in

---

21 Tacit knowledge is expressed in skills that are difficult to verbalise and transmit: for example, riding a bicycle or playing a musical instrument. See Michael Polanyi, *Personal Knowledge: Towards a Post-Critical Philosophy*, Chicago: University of Chicago Press, 1958.

22 Gilbert Ryle, 'Knowing How and Knowing That: The Presidential Address', in *Proceedings of the Aristotelian Society*, vol. 46, Aristotelian Society, New York: Wiley, 1945, 1–16. See also Hayek, 'The Primacy of the Abstract', in Koestler and Smythies, *Beyond Reductionism*.

23 Friedrich Hayek, 'Rules, Perception, and Intelligibility', *Proceedings of the British Academy* 48 (1962), reprinted in Friedrich Hayek, *Studies in Philosophy, Politics, and Economics*, London: Routledge, 1967, 44–5.

24 Hayek, 'Rules', 55.

the making, of embryonic abstractions to be perfected.[25] Thanks to these studies, Hayek was able to declare that unconscious behaviours also possess the power to make habits, norms, and abstractions. Along similar lines, Mirowski and Nik-Khah comment that 'for Hayek, it was rationality that was largely unconscious ... Knowledge here was no longer like entropy or pixie dust; now it resembled a great submerged iceberg, nine-tenths of it invisible.'[26] Although captivating, the analogy of submerged rationality is not an accurate picture of Hayek's position. Reversing the usual topology of the mind, Hayek suggested that tacit knowledge is not subconscious but rather 'supra-conscious' or 'meta-conscious'.[27] He stressed the existence of meta-conscious rules that are as abstract as conscious ones:

> While we are clearly often not aware of mental processes because they have not yet risen to the level of consciousness but proceed on what are (both physiologically and psychologically) lower levels, there is no reason why the conscious level should be the highest level, and many grounds which make it probable that, to be conscious, processes must be guided by a supra-conscious order which cannot be the object of its own representations. Mental events may thus be unconscious and uncommunicable because they proceed on too high a level as well as because they proceed on too low a level.[28]

What escapes Hayek's assessment is that this decentralised and unconscious rationality is to be found not only in markets but also in other forms of human organisation and cooperation. Marx, for example, recognised the division of labour in workshops and manufactories as a form of spontaneous and unconscious rationality.[29] Capital, according to Marx, does not just exploit workers individually but does so through the social cooperation that is augmented by the division of

---

25 Kurt Goldstein and Martin Scheerer, 'Abstract and Concrete Behavior: An Experimental Study with Special Tests', *Psychological Monographs* 53, no. 2 (1941). Goldstein's model of 'abstract attitude' or 'categorical behaviour', however, changed significantly in the 1960s.
26 Mirowski and Nik-Khah, *The Knowledge*, 68–9.
27 Hayek, 'Rules', 61.
28 Ibid.
29 See Matteo Pasquinelli, 'On the Origins of Marx's General Intellect', *Radical Philosophy* 2, no. 6 (Winter 2019): 43–56.

labour and machinery. As we saw in chapter 4, Marx assigned the power of the division of labour to the figure of the collective worker (*Gesamtarbeiter*), which is distinct from the sum of individual tasks; similarly, Hayek saw the market as a spontaneous form of self-organisation that is more than the mere sum of its individual exchanges. The difference between the two is that Marx, following Charles Babbage's lead, was aware that the spontaneous rationality of labour could be captured by the factory system and technological innovation, while Hayek assumed that the capture of the rationality of the market by a technical or institutional apparatus would be impossible and, if ever possible, illiberal. Hayek could not forecast that at the turn of the coming century, digital networks and large data centres, employing the very artificial neural networks discussed by cyberneticians, would be able to trace and compute social behaviours and collective rationality in real time, inaugurating a highly effective regime of knowledge extractivism on a global scale.

The faculty of classification; or, What is a pattern?

Throughout his career, Hayek defined classification as the main faculty of the mind in its interactions with the world and in the generation of new ideas (including those 'ideas' most crucial to economists: commodity prices). In their 1947 paper, McCulloch and Pitts already theorised artificial neural networks for 'the perception of auditory and visual forms', but Hayek's 1952 book *The Sensory Order* was the first systematic treatment of connectionism and classification as a general faculty of the mind. Even today, Hayek's account of classification remains a valid introduction to the definition of classifier algorithms in machine learning:

> The phenomena with which we are here concerned are commonly discussed in psychology under the heading of 'discrimination'. This term is somewhat misleading because it suggests a sort of 'recognition' of physical differences between the events which it discriminates, while we are concerned with a process which *creates* the distinctions in question. The same is true of most of the other available words which might be used, such as 'to sort out', 'to differentiate', or 'to

classify'. The only appropriate term which is tolerably free from misleading connotations would appear to be 'grouping'. For the purposes of the following discussion it will nevertheless be convenient to adopt the term 'to classify' with its corresponding nouns 'classes' and 'classification' in a special technical meaning . . . By 'classification' we shall mean a process in which on each occasion on which a certain recurring event happens it produces the same specific effect . . . All the different events which whenever they occur produce the same effect will be said to be events of the same class, and the fact that every one of them produces the same effect will be the sole criterion which makes them members of the same class.[30]

The above passage is followed in *The Sensory Order* by Hayek's speculations about the possibility of machines embodying this principle of classification. Hayek provided examples of analogue machines that, in their simplicity, can help illustrate the basic statistical logic of early artificial neural networks such as Rosenblatt's perceptron:

We may conceive of a machine constructed for the purpose of performing simple processes of classification of this kind. We can, for instance, imagine a machine which 'sorts out' balls of various size which are placed into it by distributing them between different receptacles . . . Another kind of machine performing this simplest kind of classification might be conceived as in a similar fashion sorting out individual signals arriving through any one of a large number of wires or tubes. We shall regard here any signal arriving through one particular wire or tube as the same recurring event which will always lead to the same action of the machine. The machine would respond similarly also to signals arriving through some different tubes or wires, and any such group to which the machine responded in the same manner would be regarded as events of the same class. Such a machine would act like a simplified telephone exchange in which each of a number of incoming wires was permanently connected with, say a particular bell, so that any signal coming in on any one of these wires would ring that bell. All the wires connected with any one bell would then carry signals belonging to the same class. An actual instance of a

---

30 Hayek, *The Sensory Order*, 48.

machine of this kind is provided by certain statistical machines for sorting cards on which punched holes represent statistical data.[31]

What this mechanical analogy helps illuminate is that, for Hayek, the mind's construction of classes (concepts, categories, patterns, prices, etc.) is not the mere grouping of perceptions and mental events that appear similar. Hayek claimed that the human mind defines classes not only by recognising similarities but often by *establishing* such similarities (also among arbitrary elements). This means that, for Hayek (as for the cyberneticians), the establishment of a class is a pragmatic gesture rather than an abstract one, much like the acquisition of an individual habit or social convention by repetition. For Hayek, different perceptual events are recognised as part of the same class whenever they trigger, in all their instances, the same effect in the nervous system or as motor response: that is, the same perceptual pattern must produce the same conscious idea and/or the same motor pattern.

Within the notion of class, Hayek included perceptual and aesthetical categories such as Gestalt and pattern, but also ethical and political ones such as habit and norm. Gestalt theory had registered a profound influence on Hayek, to the extent that his theoretical framework can be considered the translation of Gestalt principles into the economic and social field.[32] In German literature and science, the notion of Gestalt (or perceptual configuration) had played a central role since the eighteenth century, from Goethe to Mach, before being canonised in the Gestalt school's psychology of perception. As seen in the previous chapter, at the 1948 Hixon symposium, cyberneticians questioned Gestalt perception as a unique faculty of the human and advocated its mechanisation under techniques such as McCulloch and Pitts's artificial neural networks for pattern recognition. In fact, the more technical English term 'pattern' gradually replaced the German word *Gestalt*, which was imported to the United States by the diaspora of scholars fleeing Nazism.[33]

---

31 Ibid.

32 See also Nicolò De Vecchi, 'The Place of Gestalt Psychology in the Making of Hayek's Thought', *History of Political Economy* 35, no. 1 (2003): 135–62.

33 Within the English tradition, Hayek mentioned N. R. Hanson, *Patterns of Discovery*, Cambridge: Cambridge University Press, 1958; G. H. Hardy, *Mathematician's Apology*, Cambridge: Cambridge University Press, 1941, 14: 'a mathematician, like a painter or poet, is a maker of patterns.'

However, it was thanks to Gestalt theory and not cybernetics that Hayek was able to extend the definitions of class and pattern to the economic field. Already in *Sensory Order* he expanded the understanding of pattern beyond the visual sphere and in so doing covered, respectively, 'patterns within the brain', 'topological patterns', 'patterns of movements', 'temporal patterns', 'patterns of behavior', 'patterns of motor responses', 'patterns of attitude or dispositions', 'patterns of nervous impulses', and so on. He developed a large repertoire of the notion of pattern that included form, template, *Schablone*, mould, schemata, abstraction, norm, habit, disposition, arrangement, rule, and inference. However, it was first with 'The Theory of Complex Phenomena' (1961) that Hayek began to use the prescient moniker 'pattern recognition' to define classification.[34] Probably Hayek's most visionary passages are those in which mathematical equations describe multidimensional patterns (which is in fact what the equations of artificial neural networks compute with differential calculus).[35] For example, he writes:

> Many of the patterns of nature we can discover only *after* they have been constructed by our mind. The systematic construction of such new patterns is the business of mathematics. The role which geometry plays in this respect with regard to some visual patterns is merely the most familiar instance of this. The great strength of mathematics is that it enables us to describe abstract patterns which cannot be perceived by our senses, and to state the common properties of hierarchies or classes of patterns of a highly abstract character. Every algebraic equation or set of such equations defines in this sense a class of patterns, with the individual manifestation of this kind of pattern being particularized as we substitute definite values for the variables.[36]

Like other modern philosophers, Hayek made no distinction between the ability to invent a class and to change behaviour: the constitution of

---

34 Friedrich Hayek, 'The Theory of Complex Phenomena', in *The Critical Approach and Philosophy: Essays in Honor of K. R. Popper*, ed. M. Bunge, New York: The Free Press, 1964. The term 'pattern recognition' was popularised in Oliver G. Selfridge and Ulric Neisser, 'Pattern Recognition by Machine', *Scientific American* 203, no. 2 (1960): 60–9.

35 Later on, Hayek wrote also about 'patterns in multidimensional space'. Hayek, 'Rules', 53.

36 Hayek, 'Theory of Complex Phenomena'.

habits and norms follow the same logic of the constitution of ideas. Hayek extended, in this way, the faculty of constructing classes and patterns to praxis and social behaviours:

> People do behave in the same manner towards things, not because these things are identical in a physical sense, but because they have learnt to classify them as belonging to the same group, because they can put them to the same use or expect from them what to the people concerned is an equivalent effect.[37]

Nevertheless, what is crucial for any epistemology is not the definition of knowledge per se but the capacity for its invention. How does a mind invent new ideas? Hayek not only had to offer a definition of classification or pattern recognition but also had to clarify how new classes and patterns are made. For Hayek, human beings continuously make and unmake classes and patterns in their everyday activities. Specifically, the disruption of traditional and familiar classes through which reality is perceived and the reconstitution of new ones within unexpected constellations should be considered the modus operandi of science (against scientism and the 'engineering type of mind'):[38]

> The idea that science breaks up and replaces the system of classification which our sense qualities represent is less familiar, yet this is precisely what Science does . . . This process of re-classifying 'objects' which our senses have already classified in one way, of substituting for the 'secondary' qualities in which our senses arrange external stimuli a new classification based on consciously established relations between classes of events is, perhaps, the most characteristic aspect of the procedure of the natural sciences. The whole history of modern Science proves to be a process of progressive emancipation from the innate classification of the external stimuli till in the end they completely disappear.[39]

---

37 Hayek, 'Scientism and the Study of Society', 277.
38 Ibid., 269.
39 Ibid., 271–2.

Given the synthesis of psychology, mathematics, cybernetics, sociology, and the philosophy of science in his theory of connectionism, Hayek can truly be defined as the economist of pattern recognition, or better, the economist that turned pattern recognition into a market principle of neoliberalism.

Neural networks as a model of the mind

How is a set of different stimuli associated with the same class – that is, recognised as a recurrent pattern? What is the cerebral process that makes classification possible? Hayek's connectionism provided an empirical explanation for the relation between perception and cognition. Influenced by McCulloch and Pitts's idea of neural networks, Hayek simplified cognition as a simple act of decision (rather than intuition, or *Einsicht*, as in the Gestalt school).[40] In McCulloch and Pitts's model, a structure of progressive layers of nodes (made of multiple neurons or switches) filters a large input into a single binary output (a single neuron or switch) that decides if the group of input stimuli belongs to a given class or not. The solution is quite elegant: one node computes a large input into a simple binary output to signify 'yes' or 'no'. As in the modality of supervised machine learning, the end node is assigned to a given class by a convention (for instance, to the label 'apple'). It is said that the model is not isomorphic, meaning that none of its parts resembles the knowledge it interprets: there is no localised area of the network that memorises, for instance, the general form of the apple in its recognisable proportions.[41] The correct classification of stimuli depends on the overall behaviour of the computing structure.

Hayek's connectionism, however, did not advocate for a computational theory of the mind. It would be no mistake to call his theory *Gestalt connectionism* to distinguish it from McCulloch and Pitts's

---

40 On the notion of insight in the Gestalt school, see David Bates, 'Creating Insight: Gestalt Theory and the Early Computer', in *Genesis Redux: Essays in the History and Philosophy of Artificial Life*, ed. Jessica Riskin, Chicago: University of Chicago Press, 2007, 237–60.

41 For the saga of model thinking in the history of AI, see also Jean-Pierre Dupuy, *The Mechanization of the Mind: On the Origins of Cognitive Science*, Cambridge, MA: MIT Press, 2000.

logical connectionism and Rosenblatt's *statistical connectionism*. Hayek argued that the mind (which in his view was a mental order, a self-organised network of entities such as neurons) can only provide a *model* rather than a representation of the world (a sensory order, made of relations among qualia). Hayek wrote that 'what we call mind is thus a particular order of a set of events taking place in some organism and in some manner related to but not identical with, the physical order of events in the environment'.[42] In 1945, cyberneticians Arturo Rosenblueth and Norbert Wiener framed model-making in similar terms:

> Partial models, imperfect as they may be, are the only means developed by science for understanding the universe. This statement does not imply an attitude of defeatism but the recognition that the main tool of science is the human mind and that the human mind is finite.[43]

The construction of a model is the implementation of a given environment within the internal parameters and constraints of another environment, yet in the process of translation some elements are dispersed, approximated, and distorted. Hayek also acknowledged that a mental order is a partial, often false, interpretation of reality:

> We have seen that the classification of the stimuli performed by our senses will be based on a system of acquired connexions which reproduce, in a partial and imperfect manner, relations existing between the corresponding physical stimuli. The 'model' of the physical world which is thus formed will give only a very distorted reproduction of the relationships existing in that world; and the classification of these events by our senses will often prove to be false, that is, give rise to expectations which will not be borne out by events.[44]

It is telling that, after Babbage, yet another political economist is to be found at a watershed in the history of computing: Babbage proposed computation as the automation of mental labour in the industrial

---

42 Hayek, *The Sensory Order*, 16.
43 Arturo Rosenblueth and Norbert Wiener, 'The Role of Models in Science', *Philosophy of Science* 12, no. 4 (1945): 316–21.
44 Hayek, *The Sensory Order*, 145.

process, while Hayek maintained that computation of market transactions would be impossible and, in any case, detrimental to the market autonomy itself. The theoretical difference and historical gap between Babbage and Hayek mirrors the difference between symbolic and connectionist AI, between an idea of cognition based on representation and one based on modelling. Babbage's project to automate mental labour as hand calculation unfolded into the Turing machine and the deductive algorithms of symbolic AI: numerical manipulation became symbol manipulation, leaving no space for interpretation of meaning and capacity of adaptation. Whereas Babbage's computation was born through following a drive to exactitude to fix errors in logarithmic tables, a flexible and adaptive epistemology is found in connectionism (including in Hayek's variant). After Hayek and von Neumann, Rosenblatt stressed that his neural network perceptron was a simplification and exaggeration of specific traits of the human minds without claiming to be the ultimate paradigm of intelligence.[45]

The market as a model of neural networks

In addition to the theory of pattern recognition, Hayek is acknowledged for having employed, *ante litteram*, a technical definition of information. His 1945 essay 'The Use of Knowledge in Society' anticipated Shannon's 1948 mathematical theory of communication, providing an operative definition of information as units of communication – more precisely, in this case, as 'price signals'. Hayek is recognised also for describing the market as a computer – or, in the language of the time, as a sort of distributed telegraph network, 'a kind of machinery for registering change, or a system of telecommunications' (it must be noted that at the time the computer was not yet a common technology):

---

45 'Perceptrons are not intended to serve as detailed copies of any actual nervous system. They are simplified networks, designed to permit the study of lawful relationships between the organization of a nerve net, the organization of its environment, and the "psychological" performances of which the network is capable. Perceptrons might actually correspond to parts of more extended networks in biological systems; in this case, the results obtained will be directly applicable. More likely, they represent extreme simplifications of the central nervous system, in which some properties are exaggerated, others suppressed.' Rosenblatt, *Neurodynamics*, 28.

We must look at the price system as such a mechanism for communicating information if we want to understand its real function, a function which, of course, it fulfils less perfectly as prices grow more rigid... The most significant fact about this system is the economy of knowledge with which it operates, or how little the individual participants need to know in order to be able to take the right action. In abbreviated form, by a kind of symbol, only the most essential information is passed on and passed on only to those concerned. It is more than a metaphor to describe the price system as a kind of machinery for registering change, or a system of telecommunications which enables individual producers to watch merely the movement of a few pointers, as an engineer might watch the hands of a few dials, in order to adjust their activities to changes of which they may never know more than is reflected in the price movement.[46]

Contrary to the hubris of cyberneticians for full automation, Hayek asserted that the magnitude of the market's complexity would exceed the hardware limits of any apparatus of calculation and of manageable equations. Two decades later, from the other side of the socialist calculation debate, the economist Oskar Lange countered that innovation had overcome these limitations, advocating for the use of powerful new computers in solving the mathematical problems of economics: 'So what's the trouble?', Lange replied to Hayek. 'Let us put the simultaneous equations on an electronic computer and we shall obtain the solution in less than a second.'[47] Lange understood the computer as a new instrument of knowledge that inaugurates a different perspective on the economy, as 'the computer fulfils a function which the market never was able to perform'.[48] Implicitly, Lange suggested the use of the computer as technical mediator between the troubles of market spontaneity and those of centralised planning. This particular insight of Lange has been quoted by left-accelerationist rhetoric to generically foster a public use of algorithmic planning in the age of big data against the private use of

---

46 Friedrich Hayek, 'The Use of Knowledge in Society', *American Economic Review* 35, no. 4 (1945): 519–30.

47 Oskar Lange, 'The Computer and the Market', in *Socialism, Capitalism, and Economic Growth: Essays Presented to Maurice Dobb*, ed. C. H. Feinstein, Cambridge: Cambridge University Press, 1967, 158–61.

48 Ibid., 161.

said planning by corporations; Fredric Jameson, for example, advocated for the nationalisation of the computing power of global logistics giants such as Walmart and Amazon.[49] But to what specific sort of computing technique was Lange referring? Often neglected, the following part of his argument does not mention deterministic computing but something that resembles the training process of artificial neural networks:

> The market mechanism and trial and error procedure proposed in my essay really played the role of a computing device for solving a system of simultaneous equations. The solution was found by a process of iteration which was assumed to be convergent. The iterations were based on a feedback principle operating so as to gradually eliminate deviations from equilibrium. It was envisaged that the process would operate like a servo-mechanism, which, through feedback action, automatically eliminates disturbances ... The same process can be implemented by an electronic analogue machine which simulates the iteration process implied in the *tâtonnements* [incremental approximations] of the market mechanism. Such an electronic analogue (servo-mechanism) simulates the working of the market. This statement, however, may be reversed: the market simulates the electronic analogue computer. In other words, the market may be considered as a computer sui generis which serves to solve a system of simultaneous equations. It operates like an analogue machine: a servo-mechanism based on the feedback principle. The market may be considered as one of the oldest historical devices for solving simultaneous equations. The interesting thing is that the solving mechanism operates not via a physical but via a social process. It turns out that the social processes as well may serve as a basis for the operation of feedback devices leading to the solution of equations by iteration.[50]

Along the tradition of Hayek's connectionism, Lange described the market as a social machine solving simultaneous equations by incremental approximations (*tâtonnements*), in a way similar to a learning

---

49 Leigh Phillips and Michal Rozworski, *The People's Republic of Walmart: How the World's Biggest Corporations Are Laying the Foundation for Socialism*, London: Verso, 2019. See also Fredric Jameson, *Archaeologies of the Future: The Desire Called Utopia and Other Science Fictions*, London: Verso, 2005, 153n22.

50 Lange, 'The Computer and the Market', 159.

algorithm that changes its parameters with trial-and-error adjustments. Lange's example of approximation techniques to solve market equations surely does not echo centralised socialist economies but instead, nowadays, the training algorithms of artificial neural networks (such as backpropagation and gradient descent, among others). As the two passages by Hayek and Lange have shown, in the twentieth century's economic debates, models of market and computation sometimes exchanged positions, but the real issue at stake remained the agency and autonomy of the underlying social processes.

Towards a political epistemology of neural networks

Hayek's confession that he envisioned the connectionist mind as a stock capital in a continuous exchange with the market seems to confirm, in the age of AI, the seductive theory of 'real abstraction' by Marxist scholar Alfred Sohn-Rethel. In his 1970 book *Intellectual and Manual Labour*, Sohn-Rethel sketched a 'critique of epistemology' that posited the commodity form as the origin of abstract thinking itself. Sohn-Rethel argued that the exchange of goods in antiquity mediated by money would have been the first instance of abstract thought such as philosophy, since money, like philosophy, instituted a principle of abstract equivalence between material things. A commodity that is exchanged with another is, for Sohn-Rethel, a paradigmatic example of 'real abstraction' – that is, an abstraction expressed by the means of a thing. This happens even when the act of exchange is unconscious (here both Marx and Hayek would agree). In this way, the abstraction of market exchange preceded and influenced the evolution of philosophical and scientific 'conscious' abstractions.

Sohn-Rethel was convinced that the general ideas of philosophy and analytical mathematics historically emerged when the first coined money (made of *elektron*, a naturally occurring alloy of gold and silver that was abundant in Asia Minor) started to circulate as a stable general equivalent in the ancient Greek colonies.[51] According to his narrative,

---

[51] Marc Shell has noted the ironical coincidence that is contained in the expression 'electronic money' that happens to completely dematerialise the original valuable substance. See Marc Shell, *Money, Language, and Thought: Literary and Philosophical Economies from the Medieval to the Modern Era*, Berkeley: University of California Press, 1982.

once money was liberated from the control of the despot, its numeric form galvanised philosophy as the first form of secular abstraction (religion and mythology being regimes of abstraction already in operation). A few generations after *elektron* coins entered circulation and boosted commerce, the Greek colonies witnessed the first generation of the canonical Western philosophers, including Thales, Anaximander, and Anaximenes. Sohn-Rethel argued that the notions of identity, substance, divisibility, and infinity typical of the pre-Socratic philosophers mirrored the same properties that had to be measured in the new metallic medium of commerce. For him, however, secular thinking was born as a conscious and critical reaction to the ills that money brought to Greek society.

Reducing the genesis of symbolic forms only to the monetary general equivalent can open all too easily onto fatalistic readings of the pernicious influence of capitalism on the mental order, creating a state of affairs wherein it would be difficult, if not impossible, to think outside the logic of capital. In his account of the emergence of conceptual tools, Sohn-Rethel stressed only the influence of the sphere of circulation (mercantile exchange) and thus minimised the sphere of production (the social division of labour). In so doing, he overlooked the activity of reflection of labour through tools and language, which, according to other materialist epistemologies (including those of Jean Piaget and Peter Damerow) gave rise to mathematical abstractions long before the emergence of mercantile exchange.[52] In other words, the real abstraction of the social division of labour predates the real abstraction of monetary exchange and wage labour: abstract thought existed in societies where money was not circulating but the division of labour and, in particular, slavery were enforced. Hayek would have been comfortable seeing the discipline of philosophy as a mirror of the market abstractions with no reference to the potential autonomy of labour and tool-making. If Hayek's sophisticated connectionism is, then, but a sublimated version of the 'market rationality', what would an alternative epistemology of neural networks, that would not echo the neoliberal mind, look like?

In the *Grundrisse*, Marx provided a critique of Hegel's epistemology that can also be extended to Hayek's mercantile epistemology. In the

---

52 For an extensive overview of Jean Piaget and Peter Damerow's epistemologies, see Jürgen Renn, *The Evolution of Knowledge: Rethinking Science for the Anthropocene*, Princeton, NJ: Princeton University Press, 2020.

introduction to that work (written in 1857, a decade before *Capital*), Marx described the dialectics of abstract and concrete ideas as 'the method of political economy', in this way synthesising German idealism and British political economy. Questioning the given categories of everyday language, as Hegel himself proposed in the *Phenomenology of Spirit*, Marx stressed that a familiar expression such as 'labour' is the result of the long combination of different abstractions rather than a simple and originary notion from which reflection should start.[53] According to Marx, the 'scientifically correct method' starts from decomposing an idea (*Vorstellung*) into simpler concepts (*Begriff*) and then moving again from these simple concepts to recompose the whole 'as a rich totality of many determinations and relations'.[54] Hayek's description of the scientific method as the making and unmaking of the abstractions (classes, patterns, etc.) through which reality is perceived appears not dissimilar from that of Marx, though their political extrapolations obviously diverge. The creation of new ideas is, for Hayek, a subjective affair, an exercise of individual freedom, while, for Marx, it is influenced by the social relations of production and is often organic to the logic of capital. Marx took the example of labour, which appears to be an old, familiar, and simple category, which modern capitalism has transformed into an abstraction. According to Marx, in fact, industrial capitalism emerged via the imposition of 'abstract labour' – that is, labour indifferent to the specificity of 'concrete labour', labour that is transformed into commodity, into a general equivalent of labour that any worker can perform.[55] Unlike preindustrial concrete labour, abstract labour is measured in abstract time units, and workers are paid proportionally to such units.[56] Historically, the working class in its modern sense was constituted, as a new political subject, by the imposition of the general equivalent of abstract labour during the industrial age.[57]

Unlike Hayek, Marx questioned the political genealogy of the categories of economic thought. For him, the categories of thought

---

53 Karl Marx, *Grundrisse: Foundations of the Critique of Political Economy*, trans. Martin Nicolaus, London: Penguin, 1993, 101.
54 Ibid., 101.
55 Ibid., 296.
56 Labour represented in the value of a commodity is abstract labour, which is measured on the basis of socially necessary labour-time.
57 Ibid., 103–5.

– specifically that of labour – are not neutral and are, rather, intrinsic to the capitalist logic. They thus contribute to a certain normalisation, control, and exploitation of society. However, unless one is indulging in political fatalism, one must recognise that the faculty of abstraction has never been an exclusive attribute of power only. To contest abstract labour in a capitalist sense, one should consider that the faculty of abstraction belongs to the human mind in its dialectical relation with the world, with tools and techniques, not just to a sovereign apparatus, capitalist or otherwise. As political philosophers Michael Hardt and Antonio Negri justifiably remark, 'Abstraction is essential to both the functioning of capital and the critique of it.'[58] Any abstraction, any classification, is the result of a social division of labour, of contradictions and conflicts that are generative of knowledge. Similarly, Hayek's neural networks and artificial neural networks in general remain an extension of this very social division of abstract labour.

---

58 Michael Hardt and Antonio Negri, *Commonwealth*, Cambridge, MA: Harvard University Press, 2009, 127.

9

The Invention of the Perceptron

Our success in developing the perceptron means that for the first time a non-biological object will achieve an organization of its external environment in a meaningful way . . . My colleague [Marshall Yovits] disapproves of all the loose talk one hears nowadays about mechanical brains. He prefers to call our machine a self-organizing system, but, between you and me, that's precisely what any brain is.

Frank Rosenblatt, interview by the
New Yorker, 6 December 1958[1]

The role of experimentation becomes increasingly important as the systems to be considered grow in complexity, while the amount that can be accomplished by purely logical reasoning falls increasingly short of a complete understanding of the system's performance. This does not mean that an abandonment of theoretical analysis is advocated but, rather, in the spirit of Galileo, that theory must be matched by experiment at all times, and that from the interaction of theory and experiment will emerge the knowledge of the proper steps which must follow.

Frank Rosenblatt, 'Analytic Techniques
for the Study of Neural Nets', 1964[2]

---

1 'Rival', *New Yorker*, 6 December 1958, 44.
2 Frank Rosenblatt, 'Analytic Techniques for the Study of Neural Nets', *IEEE Transactions on Applications and Industry* 83, no. 74 (September 1964): 285.

Faults are defaults, yet instruments perform. A principle of science studies is that dissensus is instructive, not pathological, and that agreement is not inevitable, but to be explained. Instruments' adequate function needs comparable analysis. Then 'the big question' is how it's judged that instruments are working and, indeed, what they are.

Simon Schaffer, 'Easily Cracked', 2011[3]

'An organization of the external environment in a meaningful way'

In 1958, two years after the Dartmouth workshop on AI, the *New York Times* granted bold headlines to the project of a new 'thinking machine': the artificial neural network 'perceptron'.[4] Its inventor, the psychologist Frank Rosenblatt (at the time only thirty years old), and its sponsor, Marshall Yovits of the US Office of Naval Research, were looking for good press to justify the expenditure of taxpayer money. In praise of the invention, the newspaper cartoonishly reported that 'the Navy revealed the embryo of an electronic computer . . . that it expects will be able to walk, talk, see, write, reproduce itself and be conscious of its existence'. It was fanfare for the military, yet some of the article's exaggerations predicted the creepy future achievements of deep neural networks. For instance, the article was eerily prescient regarding face recognition and natural language processing that would emerge half a century later: 'Later Perceptrons will be able to recognize people and call out their names and instantly translate speech in one language to speech or writing in another language.'[5]

In the same year, the *New Yorker* featured more sober coverage in the form of an interview with Rosenblatt, who clarified that the perceptron was not 'a mechanical brain', as the hype claimed, but a self-organising machine that could likewise provide 'an organization of its external environment in a meaningful way'. Paraphrasing the Hebbian principle of neuroplasticity ('Neurons that fire together, wire together'), the magazine

---

3 Simon Schaffer, 'Easily Cracked: Scientific Instruments in States of Disrepair', *Isis* 102, no. 4 (2011): 707.

4 In 1949, *Time* magazine welcomed Ashby's homeostat with the same rhetoric: 'the closest thing to a synthetic brain so far designed by man.' 'The Thinking Machine', *Time*, 24 January 1949.

5 'New Navy Device Learns by Doing', *New York Times*, 8 July 1958.

gave an accurate account, for the time, of the working of an artificial neural network:

> The distinctive characteristic of the perceptron is that it interacts with its environment, forming concepts that have not been made ready for it by a human agent. Biologists claim that only biological systems see, feel, and think, but the perceptron behaves as if it saw, felt, and thought. Both computers and perceptrons have so-called memories; in the latter, however, the memory isn't a mere storehouse of deliberately selected and accumulated facts but a free, indeterminate area of association units, connecting, as nearly as possible at random, a sensory input, or eye, with a very large number of response units.
>
> If a triangle is held up to the perceptron's eye, the association units connected with the eye pick up the image of the triangle and convey it along a random succession of lines to the response units, where the image is registered. The next time the triangle is held up to the eye, its image will travel along the path already travelled by the earlier image. Significantly, once a particular response has been established, all the connections leading to that response are strengthened, and if a triangle of a different size and shape is held up to the perceptron, its image will be passed along the track that the first triangle took.[6]

What kind of artificial neural network was the perceptron? Rosenblatt struggled to explain the workings of perceptrons in simple terms and later lamented that media hype spoiled 'scientific confidence'.[7] At the Cornell Aeronautical Laboratory in Buffalo, New York, it was filed as 'Project PARA: Perceiving and Recognising Automaton'. The Navy – the main backer of the project – was essentially interested in the automation of target classification, such as the reconnaissance of enemy vessels through radar, sonar, or visual data (figs. 9.1 and 9.2).[8] Rosenblatt also planned to design, aside *photoperceptrons*, a whole class of devices operating with the same logic which included *phonoperceptrons* (to

---

6 'Rival', 44.

7 Frank Rosenblatt, *Principles of Neurodynamics: Perceptrons and the Theory of Brain Mechanisms*, Washington: Spartan Books, 1962, vii (originally published as Technical Report VG-1196-G-8, Cornell Aeronautical Laboratory, 15 March 1961).

8 See Nicholas Pryor, 'The Perceptron as an Adaptive Classification Device', US Naval Ordinance Laboratory, White Oak, Silver Spring, Maryland, November 1961.

A. AERIAL PHOTO OF SHIP IN SPECULAR SEA CLUTTER

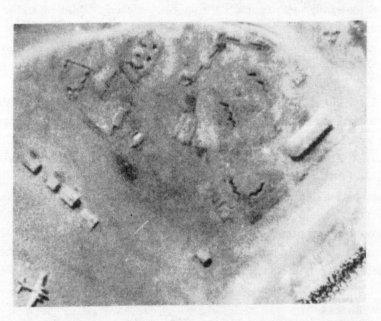

AERIAL PHOTO OF PERSONNEL TRENCHES AND OTHER AIRFIELD STRUCTURES

Figure 9.1. Examples of target classification. Albert Murray, 'Perceptron Applications in Photo Interpretation', *Photogrammetric Engineering* 27, no. 4 (1961): 633.

The Invention of the Perceptron

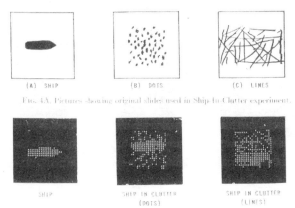

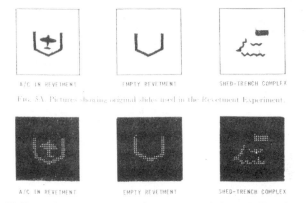

Figure 9.2. Examples of target classification. Albert Murray, 'Perceptron Applications in Photo Interpretation', *Photogrammetric Engineering* 27, no. 4 (1961): 634.

recognise words in audio communications) and *radioperceptrons* (to recognise objects in radar and sonar signals).[9] Technically speaking, the perceptron was a statistical neural network for pattern recognition – that is, a self-organising computing network for the classification of stimuli in a binary way, as briefly mentioned in previous chapters.

---

9 See Frank Rosenblatt, 'The Perceptron: A Perceiving and Recognizing Automaton (Project PARA)' Report No. 85-460-1, Cornell Aeronautical Laboratory, 2 January 1957. See also the visual examples in Albert Murray, 'Perceptron Applications in Photo Interpretation', *Photogrammetric Engineering* 27, no. 4 (1961).

Figure 9.3. Mark I Perceptron. Source: Frank Rosenblatt, *Principles of Neurodynamics: Perceptrons and the Theory of Brain Mechanisms*, Buffalo, NY: Cornell Aeronautical Laboratory, 1961, iii.

The Mark I Perceptron

The first implementation of the perceptron was a computer simulation written in the SHARE programming language and run, in 1957, on an IBM 704, one of the first commercial mainframes. In the earliest tests, the computer was fed a series of punched cards, and apparently, after fifty trials, it 'taught itself to distinguish cards marked on the left from cards marked on the right'.[10] Rosenblatt considered this proof that a more complex architecture of the perceptron could be designed to recognise more complex patterns. Shortly after the *New York Times* article, this idea took the form of a bulky piece of hardware which would be completed only in 1960: the legendary Mark I Perceptron which now rests at the Smithsonian National Museum of

---

10 Pryor, 'Perceptron as an Adaptive Classification Device', 7.

The Invention of the Perceptron

American History in Washington, DC (see fig. 9.3). The Mark I was the same digital computer that had been used by John von Neumann in the 1940s to make calculations for the Manhattan Project; however, in this implementation, it was extended by the analogue module of the Perceptron. Though it was a thousand times slower than the IBM 704, this hardware–software hybrid allowed a programmer to rewire the network by hand, which was faster than rewriting a program. The operator's manual described it in this way:

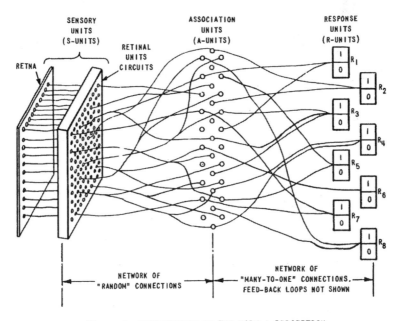

Figure 9.4. Diagram of the organisation of the Mark I Perceptron (feedback loops not shown). Frank Rosenblatt, *Mark I Perceptron Operators' Manual*. Buffalo, NY: Cornell Aeronautical Laboratory, 1960.

The Mark I Perceptron is a pattern learning and recognition device. It can learn to classify plane patterns into groups on the basis of certain geometric similarities and differences. Among the properties which it may use in its discriminations and generalizations are position in the retinal field of view, geometric form, occurrence frequency, and size. If, of the many possible bases of classification, a particular one is desired, it can generally be transferred to the perceptron by a *forced learning* session or by an *error correction training* process. If left to its own resources the perceptron can still divide up into classes the patterns presented to it, on a classification basis of its own forming. This formation process is commonly referred to as *spontaneous learning*.[11]

The Mark I Perceptron implemented a simple neural network made of three layers of units that were connected in progression: 'Sensory or S-units, Association or A-units, Response or R-units' (see fig. 9.4).[12] The input layer (also called the 'retina') was a 20-by-20-pixel camera, featuring 400 photoreceptors in total. These sensory units were randomly or topologically connected, with fixed weights, to a layer of 512 associative units.[13] The associative units were themselves connected to eight response units (R-units, or output) with weights that could be adjusted automatically (these 'weights' were analogue potentiometers that could be also controlled by hand). Like Warren McCulloch and Walter Pitts's artificial neurons, associative and response units operated according to a threshold value: they would sum up their input values and fire only if the sum was above a given threshold (see fig. 9.5).[14] The operator's manual described it as a sort of implementation of the Hebbian rule, but this is perhaps a generous interpretation:

---

11 John Hay, Ben Lynch, and David Smith, 'Mark I Perceptron Operators' Manual', Buffalo, NY: Cornell Aeronautical Laboratory, 1960, 1.
12 Ibid., 3. In the current denominations of machine learning, only the central layer counts, and the simple perceptron is termed a single-layer neural network.
13 Hay et al., 'Mark I Perceptron Operators' Manual'.
14 'All the sub-units are of a type which switch on and transmit excitation onward to the next layer only when the sum of their input excitation exceeds a certain threshold. For the Mark I, the input signals of the S-units are in the form of light, while for the rest of the units, the input signals are electrical.' Hay et al., 'Mark I Perceptron Operators' Manual'.

The A-units, however, differ from the others in that when they do switch on, the excitation which they transmit to the R-units has a value which is dependent on the comparative success which that A-unit has had in contributing to the switching of its R-unit in the past. These values form the memory of the perceptron.[15]

In this intricate tangle of wires, one has only to remember that the parameters that could be trained were the potentiometers linking A-Units with R-Units. In total, 512 x 8 = 4096 parameters. To be precise, the Mark I Perceptron was running eight simple perceptrons in parallel, each one for a dedicated pattern to be recognised. Given a retina of 20-by-20 pixels, each simple perceptron featured 400 parameters. In any case, in terms of algorithmic complexity, this was already a big number of variables to be computed by incremental approximation with the resources of the time. Today, to give an idea of the different scale of complexity, three lines of Python code suffice to run the perceptron algorithm on a desktop computer, whereas a large model such as GPT4 comprises of around 1 trillion parameters (which requires a large data centre for training and deployment).

The randomness of the initial connections and weights was crucial for Rosenblatt to demonstrate that the perceptron displayed a capacity of self-organisation, even when initialised from a chaotic state. Rosenblatt was enthusiastic about the perceptron's computing resolution, remarking: 'It is clear that with an amazingly small number of units – in contrast with the human brain's 10^{10} nerve cells – the perceptron is capable of highly sophisticated activity.'[16] The architecture of the perceptron could vary and assume different configurations with more layers and functions. It was already clear in Rosenblatt's papers that guessing the optimal configuration of the computing network was a craft of its own, forming part of the experimental practice.

Guessing a good architecture for the perceptron was only half of the problem; the other half was to design a training algorithm and error-correction technique to find the optimal value of the parameters

---

15 Hay et al., 'Mark I Perceptron Operators' Manual', 3.
16 Frank Rosenblatt, 'The Design of an Intelligent Automaton', *Research Trends*, Cornell Aeronautial Laboratory 6, no. 2 (1957): 4.

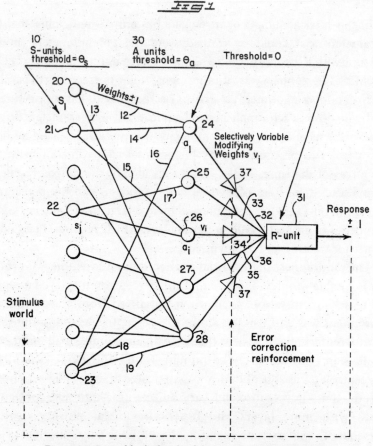

Figure 9.5. Rosenblatt's sketch of the simple perceptron in the US patent US3287649A. It was submitted in 1963 and granted in 1966. The patent expired after twenty years in 1983.

expressed by the network connections.[17] The training procedure was based on the assumption that if a solution to the classification problem

---

[17] The perceptron training algorithm was inspired by the procedure already tested in B. G. Farley and W. Clark, 'Simulation of Self-Organizing Systems by Digital Computer', *Transactions of the IRE Professional Group on Information Theory* 4 (1954): 76–84.

existed (if the set of images could be linearly separated in two groups), the parameters would converge to the optimal values in a finite number of steps. A main algorithm to train the perceptron was the following procedure of step-by-step approximation, which is basically an automated version of the technique of differential calculus:

1. Start off with a perceptron having random weights and a training set.
2. For the inputs of an example in the training set, compute the perceptron's output.
3. If the output of the perceptron does not match the output that is known to be correct for the example:
   a. If the output should have been 0 but was 1, decrease the weights that had an input of 1.
   b. If the output should have been 1 but was 0, increase the weights that had an input of 1.
4. Go to the next example in the training set and repeat steps 2–4 until the perceptron makes no more mistakes.[18]

Today's deep learning employs more refined algorithms (such as gradient descent), but the trial-and-error principle remains the same: (1) present an image to the neural network; (2) check if the output is correct; (3) if not, increase or decrease the parameters by a small value; (4) repeat the procedure until the neural network computes the correct output. Once again, the problem is finding the most efficient procedure that converges to the final result in the fewest number of steps. The design of the training algorithm is a further level of abstraction and problem-solving, which is distinct from the structure of the neural network. On this view, what many still call 'artificial intelligence' is just a technique of mathematical optimisation. This is still a case of brute-force approximation, the logic of which has become even more 'brute' in large models featuring trillions of parameters. Eventually, it is remarkable that at the heart of the most advanced techniques of 'artificial intelligence', one finds approximation procedures not dissimilar to those that constituted calculus since antiquity (see chapter 1).

---

18 Andrey Kurenkov, 'A Brief History of Neural Nets and Deep Learning', 27 September 2020, skynettoday.com.

What was the Mark I Perceptron capable of? In terms of pattern recognition, not much. It was able to distinguish a black square on the left of the visual field from one on the right, and to distinguish simple letters if they were aligned on the centre of its 20-by-20 visual matrix. Its capacity for pattern recognition (as Marvin Minsky and Seymour Papert demonstrated in their famous criticism, explained below) was primitive and restricted to continuous figures. Rosenblatt and his team were aware of its limitations, but they speculated that architectures with more layers (as deep learning eventually proved) could perform more complex tasks of recognition:

> It seems reasonable to expect that a machine similar to the Perceptron with a logical depth of three or more (obtained by two or more layers of A-Units, with each layer providing the excitation for the next) would be even more powerful than the Perceptron.[19]

It should be noted that in 1958, Rosenblatt already envisioned spatial constraints similar to the filters that grounded the idea of convolution neural networks at the origin of deep learning.[20] Specifically, in *Neurodynamics*, he mentioned David Hubel and Torsten Wiesel's study of the cat's cortex and their topological constraints, which would later inspire Kunihiko Fukushima's 'neocognitron' neural network (1980), Yann LeCun's architecture known as 'LeNet' (1989), and finally AlexNet (2012).[21]

Rosenblatt was also aware of the logical limits of statistical neural networks in imitating human intelligence, writing: 'Statistical separability alone does not provide a sufficient basis for higher order abstraction.

---

19 Pryor, 'The Perceptron as an Adaptive Classification Device', 6.

20 'It can be shown that by a suitable organization of origin points, in which the spatial distribution is constrained . . . the A-units will become particularly sensitive to the location of contours, and performance will be improved.' Frank Rosenblatt, 'The Perceptron: A Probabilistic Model for Information Storage and Organization in the Brain', *Psychological Review* 65, no. 3 (1958): 404. This idea was also picked up by Minsky and Papert in their book *Perceptrons*, discussed below. See H. D. Block, Bruce Knight, and Frank Rosenblatt, 'Analysis of a Four-Layer Series-Coupled Perceptron', *Reviews of Modern Physics* 34 (January 1962): 135–42.

21 Rosenblatt, *Neurodynamics*, 515–16. David Hubel and Torsten Wiesel, 'Receptive Fields of Single Neurons in the Cat's Striate Cortex', *Journal of Physiology* 148, no. 3 (1959): 574–91.

Some system, more advanced in principle than the perceptron, seems to be required at this point.'[22] To date, lacking a complete theory of statistical learning, artificial neural networks and deep learning are still at the epistemic stage of *experiments*. In other words, they are machines of unknown potentialities and unpredictable failures.

Brain models and experimental method

The simple perceptron was not the first artificial neural network but the first *adaptive* one – meaning it was able not just to recognise patterns but to *learn* how to recognise them (and to be rewired in different configurations in order to *learn differently*). Although its achievements were primitive, it is considered, nevertheless, the first classifier algorithm. As previously illustrated, the neural network architecture was already known: in order to demonstrate the capacity of self-organisation and adaptation of brain neurons, Rosenblatt intended to initiate the perceptron with random values. Rosenblatt then applied an error-correction algorithm to gradually adjust these values and have them converge towards an optimal equilibrium with external data, achieving in this way 'intelligence' as a spontaneous order emerging from chaos.

The hype around neural networks and self-organising theories (see chapter 6) initially caused ill feeling and envy among the 'artificial intelligentsia', especially in the circles of symbolic AI that were competing for the same military funds.[23] In order to counteract this negative response, in 1961 Rosenblatt went on to systematise his research in the lengthy monograph *Principles of Neurodynamics*, which, while little studied, remains the best source to understand the origins of artificial neural networks. However, an essay from 1964, 'Analytic Techniques for the Study of Neural Nets', better illustrates Rosenblatt's research perspective. In this later text, in order to defend the experimental nature of artificial neural networks, Rosenblatt polemically mobilised the Galilean method against the Aristotelian one, which, in his view, was still used in other

---

22   Rosenblatt, 'The Perceptron: A Probabilistic Model', 405.
23   For a parody of the competition for military funding among the different AI research groups, see Louis Fein, 'The Artificial Intelligentsia', Wescon Technical Papers, vol. 7 (11.1, part 7) (1963), 1–7.

studies of brain models. Symbolic AI theorists, for instance, believed in the possibility of encoding the mind's rules into the machine's rules straightforwardly, without experimental testing:

> For the two millennia which followed Aristotle, it was believed that the fundamental truths of nature could be revealed through the application of pure reason, that it was the philosopher, rather than the experimenter, who might discern the necessary order of nature through the sheer power of his intellect ... Then, at the beginning of the seventeenth century, the publication of Galileo's 'Discourses on Two New Sciences' gave voice for the first time to the doctrine of the experimental method. Galileo's work, advocating a clear alternative to Aristotelian rationalism, engendered a period of scientific growth and discovery in the physical sciences which has not yet run its course ... It may happen, by coincidence, that these results have application in the engineering domain, but for present purposes we propose to work not as inventors, but as discoverers, and the kind of theorizing which leads to scientific discovery is apt to be quite different from the kind of theorizing which is useful for engineering synthesis.[24]

The way in which Rosenblatt claimed the role of 'discoverer' rather than 'inventor' can be considered naive but it was somehow a defence of the experimental and scientific method against the 'engineer mentality' of many cyberneticians. Rosenblatt professed the experimental culture of artificial neural networks research in particular against the self-proving logic of symbolic AI. In the introduction to *Neurodynamics*, similarly, Rosenblatt echoed McCulloch's method of 'experimental epistemology' to assert that:

> a perceptron is first and foremost a brain model, not an invention for pattern recognition. As a brain model, its utility is in enabling us to determine the physical conditions for the emergence of various psychological properties. It is by no means a 'complete' model, and we

---

24 Frank Rosenblatt, 'Analytic Techniques for the Study of Neural Nets', *Proceedings of the American Institute of Electrical Engineers (AIEE) Joint Automatic Control Conference*, 1962, 91.

are fully aware of the simplifications which have been made from biological systems; but it is, at least, an analysable model.[25]

Siding with brain scientists rather than computer engineers was for Rosenblatt a way to contend the tentative, partial, and incomplete nature of the perceptron as an *experimental model*. In a similar fashion to Hayek, Rosenblatt maintained that a model of the brain is always an implementation, that is, a simplification and exaggeration of some of its traits, as he explained:

> Perceptrons are not intended to serve as detailed copies of any actual nervous system. They are simplified networks, designed to permit the study of lawful relationships between the organization of a nerve net, the organization of its environment, and the 'psychological' performances of which the network is capable. Perceptrons might actually correspond to parts of more extended networks in biological systems ... More likely, they represent extreme simplifications of the central nervous system, in which some properties are exaggerated, others suppressed.[26]

The perceptron was a machine constituted by numerous parameters to be adjusted in order to approximate a result. If scientific experiments usually rely on testing models of few parameters, Rosenblatt's neural network can be regarded as an experimental simulation par excellence, given the increasing number of parameters to be determined. This experimental dimension was missing from symbolic AI, whose algorithms were instead often based on the opposite assumption – that a limited number of rules could project unlimited intelligence without much acknowledgement of the critical role that implementation plays.[27] The perceptron's numerical parameters were not a *representation* of the world as in symbolic AI; they were simply relational and partial elements in the construction of a *non-isomorphic model* of the world. This feature would escalate in deep learning, with algorithmic models such as GPT4 today featuring trillions

---

25 Rosenblatt, *Neurodynamics*, viii.
26 Ibid., 28.
27 For the role of technical implementation in symbolic AI, see Stephanie Dick, 'Of Models and Machines: Implementing Bounded Rationality', *Isis* 106, no. 3 (2015): 623–34.

of parameters. In fact, despite the seeming simplicity of its architecture, a neural network requires a number of operations that exponentially increases just by adding a few layers and connections. Writing the history of computation from the point of view of algorithmic complexity – that is, the size of calculations and resource usage – statistical neural networks such as the perceptron marked a hurdle that, at the time, could not be successfully crossed due to the lack of computing power.

From symbolic logic to vector space

The first international symposium on the newly established field of AI took place in November 1958 at the National Physical Laboratory in Teddington, West London, under the title 'Mechanisation of Thought Processes'.[28] This event played a key role in the history of AI, but it has been rarely studied; here, for reasons of space, only Rosenblatt's contribution is considered. Rosenblatt took part in the symposium to clarify and defend the mathematical intuition behind the perceptron – that is, the theorem of statistical separability of data in a multidimensional space. Standing apart from the rigid computationalism of other AI scholars who attended the symposium, Rosenblatt explained that 'the mathematics of the perceptron had much more in common with 'the mathematics of particle physics', namely statistics, than with the 'mathematics of digital computers'.[29] In modelling the brain, Rosenblatt urged his colleagues to depart from the paradigm of digital computation because 'Boolean algebra, or symbolic logic, is well suited to the study of completely describable logical systems, but breaks down as soon as we attempt to apply it to systems on which complete information is not available'.[30] In favour of his thesis, Rosenblatt mobilised also the authority of von Neumann, who had passed away just the year before. As seen already in chapter 7, von Neumann, in one of his last lectures, stressed:

---

28 See also Herbert Bruderer, 'The Birth of Artificial Intelligence: First Conference on Artificial Intelligence in Paris in 1951?', in *International Communities of Invention and Innovation*, Advances in Information and Communication Technology, vol. 491, ed. Arthur Tatnall and Leslie Christopher, Berlin: Springer, 2016.

29 Frank Rosenblatt, 'Two Theorems of Statistical Separability in the Perceptron', *Symposium on the Mechanization of Thought*, National Physical Laboratory, Teddington, UK, November 1958, vol. 1, H. M. Stationery Office, London, 1959.

30 Rosenblatt, 'Two Theorems of Statistical Separability in the Perceptron', 422.

The Invention of the Perceptron

Logics and mathematics in the central nervous system . . . must structurally be essentially different from those languages to which our common experience refers . . . When we talk mathematics, we may be discussing a *secondary* language, built on the *primary* language truly used by the central nervous system. Thus the outward forms of *our* mathematics are not absolutely relevant from the point of view of evaluating what the mathematical or logical language *truly* used by the central nervous system is . . .[31]

Von Neumann argued that there is less 'logical depth' in the brain than in a computer, which may require millions of successive logical steps to imitate a simple thought process (known as the problem of 'combinatorial explosion' mentioned above). Following von Neumann, Rosenblatt similarly concluded that 'in dealing with the brain, a different kind of mathematics, primarily statistical in nature, seems to be involved [as the] brain seems to arrive at many results directly, or *intuitively*, rather than analytically'. It is clear from these passages that Rosenblatt intended to conceptualise the perceptron not as a logical but as a statistical machine – that is, as a machine quite different from the paradigm of Boolean and binary computation that was emerging in those years. The genealogy of the perceptron points to a technological lineage that is related but clearly distinct from that of the digital computer.

The invention of the perceptron, in fact, condenses influences that proceeded from diverse disciplines such as neurology, psychology, engineering, cybernetics, mathematics, and statistics. Rosenblatt's book *Neurodynamics* is the best source to evidence such a conurbation of ideas. Aside from von Neumann, in *Neurodynamics* Rosenblatt credited the contributions of Nicolas Rashevsky, McCulloch and Pitts, as well as Minsky for the idea of the logical neural network; Albert Uttley for the probabilistic model of distributed memory; Ross Ashby for the theory of self-organisation in machines; Donald Hebb and Hayek for self-reinforcement in neural pathways; and Gestalt theorists for holistic perception and distributed memory, among others.[32] But, how exactly

---

31 Quoted in Rosenblatt, 'Two Theorems of Statistical Separability'.
32 Inspired by Kurt Goldstein's neuroplasticity and faculty of abstraction (see chapter 8), Rosenblatt also made a comparison between some behaviours of the

did the perceptron constitute a breakthrough in the relation to this tradition? In a nutshell, as a technical form, the perceptron was an electromechanical computing network, but as a mathematical form, it expressed a novel trick: its adjustable parameters represented coordinates in a multidimensional vector space. This intuition has less to do with neurophysiology than with statistics. Rosenblatt's innovation was, as we will see below, to apply the statistical technique of multidimensional analysis (which had dominated US psychology in the 1950s) to image recognition. This technique has since then defined the logical form at the core of machine learning.[33]

The mathematical 'trick' to solve image recognition via multidimensional analysis can be reconstructed in this way. Each digital image in a training dataset is a two-dimensional matrix of numerical values that represent pixels. In addition to being a two-dimensional matrix, each image can also be defined as a single point in a multidimensional space whose coordinates are the values of each pixel. For example, given the resolution of the Mark I Perceptron, an image of 20-by-20 pixels is equivalent to a single point in a multidimensional space of 400 dimensions. The projection of digital images onto a multidimensional space discloses unexpected properties. In such multidimensional space, for example, points that are closer together designate similar images, while points that are further apart designate dissimilar images. Furthermore, following the progression of a value along a single dimension, images can be arranged according to a specific gradient of similarity. In such a multidimensional space, pattern recognition can be performed, then, by separating a cluster of points (which represent a class of similar images) from all the others (which represent different images). A boundary (or 'hyperplane' in technical terms) can then be drawn to separate this dataspace in two regions, in order to declare which images belong to a class and which do not. The separation of the dataspace into two regions is called binary classification (from which also the term 'classifier algorithm' derives).

---

perceptron and those of patients with brain damage. 'In its "symbolic behavior", the perceptron shows some striking similarities to Goldstein's brain-damaged patients.' Frank Rosenblatt, 'The Perceptron: A Probabilistic Model for Information Storage and Organization in the Brain', *Psychological Review* 65, no. 3 (1958): 404.

33 Other algorithms of statistical learning and multidimensional analysis such as Vladimir Vapnik's support vector machines were developed independently in the Soviet Union a few years after Rosenblatt's perceptron.

Rosenblatt's theorems of statistical separability argued that a perceptron could automate this act of classification on its own and find a hyperplane to linearly separate the vector space into two regions: one containing the images corresponding to the pattern to be 'learned', the other not. The parameters of the mathematical function of the hyperplane are the weights of the network connections. The weights of the perceptron plot the hyperplane and adjust its inclination across the hyperspace until the two clusters are perfectly separated. In the case of the simple perceptron (with 400 weights between association and output units), the hyperplane would be defined by a linear equation with 400 unknowns. The values of this equation (which are the weights of the neural network) are found by the training algorithm through the step-by-step procedure of approximation abovementioned.

The perceptron is a crucial episode in the history of cultural techniques: it entails not just a process of digitisation of the picture plane into a two-dimensional numerical matrix but its vectorisation into a statistical matrix of multiple dimensions. With this method, the human faculty of image recognition was translated and reduced to a problem of mathematical optimisation in a vector space.[34] Since then, however, its influence has gone far beyond image recognition: vectorisation in multiple dimensions has been applied to all kinds of data and has come to represent the epistemic form of the 'intelligence' that machine learning embodies in general, which is a form of *statistical intelligence*.[35] The characteristic of 'intelligence' that is anthropomorphised in AI systems is essentially the trick of projecting data on a multidimensional space in order to perform operations of clustering, classification, and prediction. At its core, machine learning exhibits the quality of geometric and spatial 'intelligence'.

---

34 Other solutions to linearly separate a vector space exist. Support vector machines, for example, are algorithms that differ from artificial neural networks but aim to calculate similar hyperplanes in a multidimensional space.

35 For a history of statistical methods in pattern recognition, data science, and predictive analytics, see Matthew L. Jones, 'How We Became Instrumentalists (Again): Data Positivism since World War II', *Historical Studies in the Natural Sciences* 48, no. 5 (2018): 673–84.

The new vectors of mind

During the 1950s, psychometrics emerged as an influential subfield in the departments of psychology across US universities. It represented quite a reductionist turn in the study of the psyche, as it was mainly concerned with the quantification and statistical measurement of personality traits, cognitive abilities, and work skills. Eventually, it became a common practice for many students to render data from psychological tests into vectors in order to calculate similarities, covariances, and find patterns of different sorts.

Tracing the origins of the perceptron, the AI scholar Jonathan Penn has recently found that Rosenblatt already employed psychometric techniques of multidimensional analysis in his doctoral research, with the purpose of examining personality profiles. In 1953, Rosenblatt asked two hundred students of Cornell University to answer a questionnaire about their childhood using a numerical scale for each answer, pursuing the typical postulate of psychometrics that 'personalities can be classified objectively'.[36] In the tradition of the psychometrics of Alfred Binet, Lewis Terman, Charles Spearman, and especially Louis Leon Thurstone, Rosenblatt analysed the results through a method of factor analysis in order to compute the similarity between the numerical matrices of each questionnaire.[37] In this way, the twenty-five-year-old Rosenblatt intended to prove the mathematical presence of clusters of similar answers and, therefore, to demonstrate, as a psychometrician of sound faith would do, the existence of distinguishable personalities.

It was probably towards the end of his PhD that Rosenblatt noticed that the numerical matrices of cognitive tests looked identical to the numerical matrices of digital images and began to consider applying the same techniques of multidimensional analysis to visual pattern recognition. It is apparent that Rosenblatt's perceptron was computing patterns of similarity in numerical images in the same way in which psychometrics was computing patterns of similarity in the numerical matrices of

---

36 'Machine Computes Tests', *Cornel Alumni News*, 1 December 1953.

37 Louis Leon Thurstone, *The Vectors of Mind: Multiple Factor Analysis for the Isolation of Primary Traits*, Chicago: University of Chicago Press, 1935; Louis Leon Thurstone, *Multiple-Factor Analysis: A Development and Expansion of the Vectors of Mind*, Chicago: University of Chicago Press, 1947.

psychological profiles.[38] This is another example of the spurious and experimental genealogies of AI, which points, however, to a specific and intriguing modality of technological innovation in which metrics anticipates automation: Rosenblatt, in fact, repurposed the tools that were used to quantify a cognitive task for the automation of the cognitive task itself.

During his PhD, Rosenblatt had another idea that served as a precursor to the perceptron: he planned to automate statistical analysis with a new calculating machine that was then called the Electronic Profile Analyzing Computer (EPAC).[39] The journal *American Scientist* described it in this way:

> An 'idiot brain', an electronic computer that can solve only one type of problem, has been designed and built by a 25-year-old psychology student at Cornell University. The machine is helping its inventor, Frank Rosenblatt to prepare data for his Ph.D. thesis at the University. A problem that would take 15 minutes to solve with a desk computer can be solved by the machine in two seconds. Rosenblatt is testing the idea that personalities can be classified in a scientific and objective way.[40]

Predating the Mark I Perceptron, Rosenblatt's EPAC was a first experiment in the automation of multidimensional analysis, a task which was commonly assigned to human 'computers' (often women) in the psychology laboratories of the time. In the same way in which Babbage put a calculating engine in place of a human computer, it could be said that Rosenblatt put a computer in place of a statistician, shaping machine

---

38 Probably Rosenblatt was also influenced by the fact that 'pattern perception' is a common test in psychometrics. For examples of similarities between visual and non-visual data matrices, see Thurstone, *The Vectors of Mind*.

39 'The purpose of EPAC was to process data collected from a paid, six-hundred item survey of more than two-hundred Cornell undergraduates. The survey was conducted on paper and then fed into the computer. These data would be used to test the viability of Rosenblatt's k-coefficient against a novel baseline: the complex relations between personality type and familial relationships during a student's first twelve years of life. Survey questions pertained to personal relationships between a student, their parents and their siblings, as well as other permutations of that roster.' Jonathan Penn, 'Inventing Intelligence: On the History of Complex Information Processing and Artificial Intelligence in the United States in the Mid-Twentieth Century', PhD dissertation, Cambridge University, 2020, 89; Frank Rosenblatt, 'The K-Coefficient: Design and Trial Application of a New Technique for Multivariate Analysis', PhD dissertation., Cornell University, 1956.

40 'Editor Miscellany', *American Scientist* 42, no. 1 (January 1954): 32.

learning as it is understood today. During his PhD, Rosenblatt aimed to empower psychometrics with the help of a computer, but it was psychometrics, in fact, that helped to calculate the matrices of artificial neural networks and contributed to forge a new – statistical this time – model of the synthetic mind.

It is of historical relevance that the perceptron advanced the automation of statistical tools precisely in the same years when they were becoming a predominant method in US psychology. This institutionalisation of statistics in US psychology between 1940 and 1955 has been studied and confirmed also by the German psychologist Gerd Gigerenzer. In addition, Gigerenzer has noticed another critical phenomenon, the transformation of the tools of psychological analysis into a theory of the mind in its own right:

> The statisticians' conquest of new territory in psychology started in the 1940s . . . By the early 1950s, half of the psychology departments in leading American universities offered courses on Fisherian methods and had made inferential statistics a graduate program requirement. By 1955, more than 80% of the experimental articles in leading journals used inferential statistics to justify conclusions from the data . . . I therefore use 1955 as a rough date for the institutionalization of the tool in curricula, textbooks, and editorials . . . In experimental psychology, inferential statistics became almost synonymous with scientific method. Inferential statistics, in turn, provided a large part of the new concepts for mental processes that have fueled the so-called cognitive revolution since the 1960s. Theories of cognition were cleansed of terms such as restructuring and insight, and the new mind has come to be portrayed as drawing random samples from nervous fibers, computing probabilities, calculating analyses of variance (ANOVA), setting decision criteria, and performing utility analyses. After the institutionalization of inferential statistics, a broad range of cognitive processes, conscious and unconscious, elementary and complex, were reinterpreted as involving 'intuitive statistics'.[41]

---

41 Gerd Gigerenzer, 'From Tools to Theories: A Heuristic of Discovery in Cognitive Psychology', *Psychological Review* 98, no. 2 (1991): 258, 255.

Gigerenzer provides a realistic periodisation that is compatible with the adoption of statistical techniques also in the artificial neural networks research. Considering Rosenblatt's PhD (1956) and his first paper on the perceptron (1957), the 1950s are indeed the years in which statistical tools of multidimensional analysis made an interdisciplinary leap and were applied to artificial neural networks and the automation of pattern recognition. It is through this path that psychometrics entered the history of AI and imparted a statistical mentality to it.

These advancements come as no surprise considering that, at the beginning of the century, psychology had already attempted to quantify human intelligence in a statistical way. Indeed, the automation of intelligence in the twentieth century was prepared by its measurement in the nineteenth century, by the establishment of a standard metrology of cognitive abilities (such as solving a puzzle or recognising a picture) rather than the study of the mind's logic. As pointed out by the historian of science Simon Schaffer:

> Since the Enlightenment, neurology, anthropology and physiology have often relied on such measures: oxygen flow, pulse rate, galvanic activity, phrenological charts, cerebral thermometry or – most pervasively – cranial capacity have all been used as markers of underlying brain activity and thus intellectual, social and moral rank. No doubt the instruments used to make such measures then become the source of neurological metaphor. But this kind of cerebral metrology embraces a wider history than that which links craniometry with more recent strategies of intelligence testing and psychometrics.[42]

To what extent could the performance of a machine be judged as 'intelligent' – that is, commensurable (measurable in the same terms) with human intelligence? Since the Turing test, machines have been judged as 'intelligent' by comparing their behaviour with social conventions. Cybernetics investigated this question in a different way, that is, by postulating a common 'mechanism' (whatever logical or

---

42 Schaffer has also noted that 'judgements that machines are intelligent have involved techniques for measuring brains' outputs. These techniques show how discretionary behaviour is connected with status of those who rely on intelligence for their social legitimacy'. Simon Schaffer, 'OK Computer', in *Ansichten der Wissenschaftsgeschichte*, ed. Michael Hagner, Frankfurt: Fischer, 2001, 393–429.

physiological) between humans and machines. But in the decades prior to cybernetics and computer science, psychometrics had already turned human intelligence into a quantifiable (and potentially computable) object. In the early twentieth century, Spearman, for instance, proposed the statistical measurement of 'general intelligence' (or g factor) as the correlation between unrelated tasks in a skill test.[43] For Spearman, these correlations mathematically demonstrated the existence of an underlying cognitive faculty that common sense would refer to as 'intelligence'.[44] Spearman's analysis was based on two factors: general intelligence (g) and specific ability (s). A few decades later, Thurstone criticised Spearman's reduction of intelligence to only two factors and proposed the consideration of multiple factors, listing up to seven intelligence attributes or 'primary mental abilities'.[45] There was enthusiasm about the flexibility of these statistical techniques which could potentially escalate the number of their dimensions and model the most complex aspects of mind and the world. In 1935, Thurstone published a book by the visionary title *The Vectors of Mind*, which aimed to provide students with an accessible introduction to multifactor analysis, moving psychology closer and closer to the mentality of statistics.[46]

Such a quantitative measure of intelligence, abstracted from social circumstances and deprived of any historical contexts, supported, however, a meritocratic social order and helped consolidate, among others, the questionable practice of measuring the intelligence quotient (IQ). These techniques were, and still are, instrumental to maintaining social hierarchies and racial segregation, and in disciplining the

---

43 'In 1904 Spearman . . . showed that aptitude tests applied to children were highly correlated. He conceived the method of factor analysis in principal components, seeking, in the vectorial space constituted by these tests, the orthogonal axes that successively explained the maximum of the variance of the cluster of points corresponding to the results for each child. Because of the correlation between the tests, the first of these axes preserved a major part of the total variance. It was a kind of average of the different tests. Spearman called it *general intelligence*, or *g factor*.' Alain Desrosières, *The Politics of Large Numbers: A History of Statistical Reasoning*, Cambridge, MA: Harvard University Press, 1998, 145.

44 Charles Spearman, 'General Intelligence: Objectively Determined and Measured', *American Journal of Psychology* 15, no. 2 (1904): 201–92.

45 Alexander Beaujean and Nicholas Benson, 'The One and the Many: Enduring Legacies of Spearman and Thurstone on Intelligence Test Score Interpretation', *Applied Measurement in Education* 32, no. 3 (2019): 198–215.

46 Thurstone, *The Vectors of Mind*.

workforce. It must be remembered that the pseudoscience of psychometrics was founded by the English statistician Francis Galton with the racist and eugenicist agenda of demonstrating a correlation between intelligence and ethnicity.[47] It is perhaps no coincidence that a system for mathematically discriminating between humans of different classes and 'races' has been subsequently used to equate humans to machines.

Spearman's g factor contributed to the reification of 'intelligence' as a new scientific 'object' that could be statistically measured. As mentioned above, Gigerenzer has also noticed a similar process of reification of a tool of inquiry into a paradigm of thought in the field of psychology – a process which he defines as 'tool-to-theory heuristics'.[48] As he noted, in the psychology of the mid-twentieth century, the 'statistical tools' of psychometrics eventually 'turned into theories of mind' in psychology. Together with Daniel Goldstein, Gigerenzer has described how the adoption of statistical tools gradually also popularised the computational metaphor of the mind, adding to its plausibility. According to them, it was specifically the influence of statistical tools such as the Neyman–Pearson theory of hypothesis testing and Roland Fisher's analysis of variance (ANOVA) that helped to consolidate the metaphor of the mind as a computer in the second half of the twentieth century.[49]

The transformation of a tool of inquiry into a model of the mind is exemplified also by the case of the perceptron, in which a statistical technique implicitly became a brain model (and, ultimately, a model of collective knowledge). The invention of statistical neural networks implied, in their construction, the 'mind as an intuitive statistician' and, conversely, also made statistics the model of the new artificial mind.[50] Statistical tools have become since then not only a model of 'intelligence' in psychology but also a model of 'artificial intelligence' in the

---

47 See Stephen Jay Gould, *The Mismeasure of Man*, New York: Norton, 1981. On AI as a technology of whiteness, see also Yarden Katz, *Artificial Whiteness: Politics and Ideology in Artificial Intelligence*, New York: Columbia University Press, 2020. On the racial subtext of the automation discourse, see Neda Atanasoski and Vora Kalindi, *Surrogate Humanity: Race, Robots, and the Politics of Technological Futures*, Durham, NC: Duke University Press, 2019.

48 Gerd Gigerenzer, 'From Tools to Theories: A Heuristic of Discovery in Cognitive Psychology', *Psychological Review* 98, no. 2 (1991): 258, 255.

49 Gerd Gigerenzer and Daniel G. Goldstein, 'Mind as Computer: Birth of a Metaphor', *Creativity Research Journal* 9, nos. 2–3 (1996): 138.

50 Ibid.

development of labour automation. Eventually, a whole statistical view of the world and society underwent a process of automation, so to speak, as it became increasingly normalised and naturalised through AI.

Hacking the vector space

In their 1969 book *Perceptrons*, Marvin Minsky and Seymour Papert demonstrated in mathematical terms that Rosenblatt's simple perceptron was unable to recognise certain patterns, questioning in this way its capacity for generalisation to other tasks of human intelligence.[51] Specifically, the book argued that certain images, once projected onto the multidimensional space, could not be linearly separated by the simple perceptron: in particular, the perceptron could not discriminate connected from disconnected figures. The theorem was illustrated with odd-shaped images that could lead the perceptron to misfire a wrong classification, and the cover of the book featured two intricate spirals that could also deceive the human eye (while they looked identical at first sight, one was continuous and the other composed of two distinguished spirals). In logical terms, the theorem explained that a perceptron reduced to only two input neurons could 'learn' the AND, OR, and NOT logic functions but not the more complex XOR (exclusive-OR).[52]

For the first time, the vector space of an artificial neural network was 'hacked' and its vulnerability exposed. Minsky and Papert's conclusions were true only for the simplest class of perceptrons (which featured only one layer of neurons), but they were received as valid for all configurations of artificial neural networks and had a devastating effect on the whole research field. This initiated the first 'winter of AI' – which was in fact the imposition of the 'MIT winter' on other research communities. Minsky and Papert's attitude was somewhat baronial and recalcitrant:

---

51 Marvin Minsky and Seymour Papert, *Perceptrons: An Introduction to Computational Geometry*, Cambridge, MA: MIT Press, 1969.

52 Already in 1960, Minsky was expressing loud scepticism about randomly connected neural networks and the logic of self-organising 'cell assemblies'. See Marvin Minsky and Oliver Selfridge, *Learning in Random Nets*, vol. 46, MIT Lincoln Laboratory, 1960. Minsky's PhD was about neural networks: 'Neural Nets and the Brain Model Problem', PhD dissertation, Princeton University, 1954.

they manifestly sought to divert military funding back to MIT – not exactly a needy institution – and to demonstrate that artificial neural networks did not constitute true 'artificial intelligence', as other techniques would reveal the virtuous path towards this goal. It should be remembered that already in his 1961 monograph *Neurodynamics*, Rosenblatt proposed different configurations of 'multi-layer perceptrons' that could overcome these limitations, but a convergence theorem could not be proven, and an efficient training algorithm (such as gradient descent) was not yet known. As the computer scientist Richard Forsyth has noted, however, only two years after the publication of *Perceptron*, in 1971, it was proven that 'a simple Mark I Perceptron, modified to incorporate an expansion recorder, could be taught to solve the exclusive-OR problem' but 'it had no effect on the widespread belief among computer scientists that neurocomputing was something that had been tried and had failed'.[53]

Among other aspects, Minsky and Papert noticed (as also had Rosenblatt) that artificial neural networks are not able to distinguish well between figure and ground: in their computation of the visual field, each point gains somehow the same priority – which is not the case with human vision. This happens because artificial neural networks have no 'concept' of figure and ground, which they replace with a statistical distribution of correlations (while the figure–ground relation implies a model of causation). The problem has not disappeared with deep learning: it has been discovered that large convolutional neural networks such as AlexNet, GoogleNet, and ResNet-50 are still biased towards texture in relation to shape. It may happen that they discriminate between the images of an elephant and a cat, for instance, not according to their shape but according to the texture of their skin and fur, respectively. The texture-over-shape bias occurs because even convolutional neural networks (which are specifically designed to extract edges, features, and details) still compute a statistical distribution of the whole data and not only of its 'meaningful' parts (as the human mind does, according to the Gestalt school). This was all the more true in the case of Rosenblatt's simple perceptron back then, but this problem of resolution in the rendering of common

---

53 Richard Forsyth, 'The Strange Story of the Perceptron', *Artificial Intelligence Review* 4, no. 2 (1990): 147–55.

knowledge extends probably also to large foundation models such as the contemporary GPT.[54]

One can argue that Minsky and Papert conceived the first adversarial method for hacking an 'intelligent machine' and designed the corresponding 'adversarial patches', as they are called today – the doctored pictures that can fool deep neural networks for image recognition.[55] As a hack, it was quite successful in that it managed to derail military funding and neural networks research until the late 1980s. Beyond the issues at stake in the controversy, Minsky and Papert nonetheless contributed to a critique of the knowledge paradigm that artificial neural networks embody and to the divulgation of the limitations of multidimensional modelling.

However, there is a tendency in the AI community, including in critical AI studies, to take sides in the 'perceptrons' controversy, mobilising viewpoints and philosophical traditions that would, alternatively, justify either symbolic or connectionist AI as the more rational or progressive paradigm, or as the more capable of causal thinking. Other readings conflate both schools under the same instrumentalist agenda of the military and its power genealogy. This book has proposed a different approach, namely to study and evaluate these AI lineages from the (externalist) perspective of labour automation, rather than as an (internalist) problem of computational logic, task performance, and human likeness. Neither deductive algorithms nor statistical techniques excel in mimicking human intelligence, because there is no inner logic to discover in human intelligence. Human cognition and machine tasks can be studied and compared because intelligence, whether 'natural' or 'artificial', is extroverted, contextual, and situated by constitution. Machines can be perceived as 'thinking' because they mimic the theatre of the human.

The adoption of statistical tools by machine learning is a counterproof, in all its controversial legacy, that the master algorithm of 'artificial general intelligence', understood as the dream of technological

---

54   The concept of 'operational image' has been recently used to stress the non-human dimension of such 'images'. Andreas Broeckmann, in particular, has proposed to define such images as 'optical calculus' also in the field of machine learning. Images, however, remain complex artefacts. They are never purely retinal, neural, technical, or logical, but always intrinsically social. Andreas Broeckmann, 'Optical Calculus', Images beyond Control Conference, FAMU, Prague, 6 November 2020.

55   Tom Brown et al., 'Adversarial Patch', NIPS Workshop, 2017, arxiv.org.

singularity and *alpha machine* by a large community of engineers and computer scientists, is precisely a statistical illusion projected by data. In other words, the master algorithm does not exist as algorithm, but only as an originary extended social form.

The social calculus of knowledge

In the 1980s, in his book *The Vision Machine*, it was the French theorist Paul Virilio who would rediscover the then little-known case of the perceptron as part of the industrial and military spectrum of projects for the 'automation of perception'. Yet these military origins should not distract from seeing the perceptron in the larger genealogy of labour automation, social control, and knowledge extractivism. Alongside the known automation of manual and mental labour, the perceptron pioneered the automation of a different kind of labour: the *labour of perception*, or *supervision*. This is the task of surveilling machines (*Maschinenarbeit* in Marx) but also workspaces and assembly lines with a clear disciplinary function when it takes place under the eye of the authority, such as masters, guards, and policemen. As the media scholar Jonathan Beller has summarised, 'to look is to labour' and has been so for a long time; but 'to look is to organise labour' as well, and the eye of the master has been doing so all along.[56] Optical media such as cinema and photography have often been involved in the automation of the gaze and the surveillance of labour in the past, and experiments of pattern recognition such as the perceptron have simply articulated these previous regimes of machine vision to a further stage.

As seen in chapter 2, the industrial age pursued the mechanisation of manual labour with tooling machines and steam engines, and, with Babbage, the mechanisation of mental labour under the form of hand calculation and symbol manipulation (which are still quite 'manual' activities, as the names indicate). In the mid-twentieth century, mainframe computers extended the automation of mental labour as calculation and symbol manipulation in state administration, large companies, and scientific research. Compared to this history, the labour of

---

56 Jonathan Beller, *The Cinematic Mode of Production: Attention Economy and the Society of the Spectacle*, Lebanon, NH: University Press of New England, 2006, 2.

supervision was mechanised in a different way. A novel aspect of the perceptron (and of pattern recognition algorithms in general) lies in the fact that a machine, for the first time, sought to automate so high a speculative faculty as the act of recognising – that is interpreting – an image, as opposed to the manipulation of symbols of given meaning. Rosenblatt defined the perceptron as a machine for the 'interpretation of the environment', arguing that the 'conceptualization of the environment is the first step towards creative thinking', and under this respect, in fact, the perceptron can be defined also as an *interpretation machine*.[57]

In today's technical jargon of machine learning, the perceptron is a *classifier* – that is, an algorithm for statistically discriminating among images and assigning them a class or category (also known as a 'label') in a given cultural taxonomy. This, perhaps the most important aspect of the classifiers, has nothing to do with their *internal logic* but with the association of their output to an *external convention* that establishes the meaning of an image or other symbol in a given culture. Gestalt theory, cybernetics, and symbolic AI each intended to identify the *internal laws* of perception, but the key feature of a classifier such as the perceptron is to record *external rules* – that is, social conventions. Ultimately, an artificial neural network is an *extroverted machine*, a machine projected towards the outside, because the interpretation of an image, for example, is always affected more by experience and external social factors than by internal physiological circuits.

A classification algorithm such as the perceptron does not automate reasoning understood as a capacity of *symbolic manipulation*, but rather as *situated knowledge* which is part of the cultural heritage of a given context. The act of image recognition or pattern classification is a specific kind of mental labour: it is a profoundly social act that mobilises, at the same time, tacit and explicit know-how, scientific and traditional taxonomies, and vernacular and technical grammars – in short, knowledge-making as a historical and often conflicting process. Although the industrial task of machine supervision can be highly codified, pattern recognition 'in the wild' remains a gesture of open interpretation rather than a strict rule-based procedure. For these reasons, a machine which

---

57 Rosenblatt, 'Two Theorems', 423–4. For a contemporary understanding of interpretation machine, see Larry Lohmann, 'Interpretation Machines: Contradictions of Artificial Intelligence in 21st-Century Capitalism', *Socialist Register* 57 (2020).

is designed to automate such an epistemic mess (see the project of self-driving cars) encounters, then as now, great difficulties. Moreover, the recent debates on gender, class, and racial bias in machine learning systems for face recognition reminds us what semioticians, philosophers of language, and art historians have always known: that image interpretation is an act that bears unresolvable political implications. In this regard, critical AI scholars Michael Castelle and Tyler Reigeluth have proposed to compare machine learning to the theory of learning as a social process by the Soviet psychologist Lev Vygotsky.[58] The semiotic structure of the classifier, which is basically an *imitation machine*, confirms what Vygotsky argued: that there is no inner logic to discover in intelligence, because intelligence is a social process by constitution.

In conclusion, Rosenblatt's initial experiment to automate pattern recognition with a small matrix of 400 dimensions resulted, after the design of convolutional neural networks in the 1980s and the rise of deep learning in the 2010s, in the algorithmic modelling of vast inventories of spontaneous knowledge, mass communication, and cultural heritage. The perceptron was an experiment of visual pattern recognition that thereafter was extended to the analysis of non-visual data, into a novel 'pattern recognition' across datasets of cultural, social, and scientific kind. In the age of deep learning, the architecture of the multi-layered perceptron ended up being not a model of the biological brain but of the collective mind, eventually expressing its original ontology shaped by psychometrics. In fact, probably the most crucial moment in the history of AI is when, with Rosenblatt's perceptron, artificial neural networks inherited the techniques of multidimensional analysis from psychometrics and statistics. This made possible not only pattern recognition but also the computation of data of much-higher dimensions – a feature that would come to be key, half a century later, in the age of 'big data'. As is well known, the unfortunate term 'big data' does refer to data that are not only vast in size but diverse in terms of typology – rendered in statistics as 'dimensions'. Nowadays, companies such as Google, Amazon, Facebook, and Twitter, for example, collect data that define an

---

58 Michael Castelle and Tyler Reigeluth, 'What Type of Learning Is Machine Learning?', in *The Cultural Life of Machine Learning: An Incursion into Critical AI Studies*, ed. Michael Castelle and Jonathan Roberge, Berlin: Springer International Publishing, 2020.

extensive manifold of dimensions about their users, such as location, age, gender, nationality, language, education, job, number of contacts, along with political orientation, cultural interests, and so on. The variety of social dimensions that these platform companies can analyse is vertiginous, exceeding the imagination of any well-trained statistician. The rise of machine learning algorithms is, then, also the response to the *dimensionality explosion* of social data rather than simply to an issue of information overload.

Eventually, in the past decade, machine learning has grown into an extensive *algorithmic modelling of collective knowledge*, a 'social calculus' that aims to encode individual behaviours, community life, and cultural heritage under the form of vast architectures of statistical correlations.[59] This has helped establish a monopolistic regime of knowledge extractivism on a global scale and new techniques for the automation of labour and management. As with only a few other artefacts of our epoch, AI has come to exemplify a unique concentration of power as knowledge.

---

59 For a general cartography of the AI production pipeline and its limitations, see Matteo Pasquinelli and Vladan Joler, 'The Nooscope Manifested: Artificial Intelligence as Instrument of Knowledge Extractivism', *AI and Society* 36 (2021), nooscope.ai.

Conclusion: The Automation of General Intelligence

We want to ask the right questions. How do the tools work? Who finances and builds them, and how are they used? Whom do they enrich, and whom do they impoverish? What futures do they make feasible, and which ones do they foreclose? We're not looking for answers. We're looking for logic.

Logic Magazine *manifesto, 2017*[1]

We live in the age of digital data, and in that age mathematics has become the parliament of politics. The social law has become interwoven with models, theorems and algorithms. With digital data, mathematics has become the dominant means in which human beings coordinate with technology . . . Mathematics is a human activity after all. Like any other human activity, it carries the possibilities of both emancipation and oppression.

Politically Mathematics *manifesto, 2019*[2]

Relics of bygone instruments of labour possess the same importance for the investigation of extinct economic formations of society as do fossil bones for the determination of extinct species of animals. It is not what

---

1 See 'Disruption: A Manifesto', *Logic Magazine*, issue 1, March 2017.
2 Politically Mathematics collective, 'Politically Mathematics Manifesto', 2019, politicallymath.in.

is made but how, and by what instruments of labour, that distinguishes different economic epochs. Instruments of labour not only supply a standard of the degree of development which human labour has attained, but they also indicate the social relations within which men work.

Karl Marx, *Capital*, 1867³

There will be a day in the future when current AI will be considered an archaism, one technical fossil to study among others. In the passage from *Capital* quoted above, Marx suggested a similar analogy that resonates with today's science and technology studies: in the same way in which fossil bones disclose the nature of ancient species and the ecosystems in which they lived, similarly, technical artefacts reveal the form of the society that surrounds and runs them. The analogy is relevant, I think, for all machines and also for machine learning, whose abstract models do in reality encode a concretion of social relations and collective behaviours, as this book has tried to demonstrate in reformulating the nineteenth-century labour theory of automation for the age of AI.

This book began with a simple question: What relation exists between labour, rules, and automation, i.e., the invention of new technologies? To answer this question, it has illuminated practices, machines, and algorithms from different perspectives – from the 'concrete' dimension of production and the 'abstract' dimension of disciplines such as mathematics and computer science. The concern, however, has not been to repeat the separation of the concrete and abstract domains but to see their coevolution throughout history: eventually to investigate labour, rules, and automation, dialectically, as *material abstractions*. Chapter 1 emphasised this aspect by highlighting how ancient rituals, counting tools, and 'social algorithms' all contributed to the making of mathematical ideas. To affirm, as did the introduction, that *labour is a logical activity* is not a way of abdicating to the mentality of industrial machines and corporate algorithms, but rather of recognising that human praxis expresses its own logic (an *anti-logic*, some might say) – a power of speculation and invention, before technoscience captures and alienates it.⁴

---

3 Karl Marx, *Capital*, vol. 1, London: Penguin, 1976, 286.
4 See the labour process debate: Harry Braverman, *Labor and Monopoly Capital: The Degradation of Work in the Twentieth Century*, New York: Monthly Review Press, 1974;

Conclusion: The Automation of General Intelligence

The thesis that labour has to become 'mechanical' on its own, before machinery replaces it, is an old fundamental principle that has simply been forgotten. As illustrated in part I, it dates back at least to Adam Smith's exposition in *The Wealth of Nations* (1776), which Hegel also commented upon already in his Jena lectures (1805–06). Hegel's notion of *abstract labour*, as labour that gives *form* to machinery, was already indebted to British political economy before Marx contributed his own radical critique of the concept. As seen in chapter 2, however, it fell to Charles Babbage to systematise Adam Smith's insight in a consistent *labour theory of automation*. Babbage complemented this theory with the *principle of labour calculation* (known since then as the 'Babbage principle') to indicate that the division of labour also allows the precise computation of labour costs. Part I of this book can be considered an exegesis of Babbage's two *principles of labour analysis* and their influence on the common history of political economy, automated computation, and machine intelligence. Although it may sound anachronistic, Marx's theory of automation and relative surplus-value extraction share common postulates with the first projects of machine intelligence.

Marx overturned the industrialist perspective – 'the eye of the master' – that was inherent in Babbage's principles. In *Capital*, he argued that the *social relations of production* (the division of labour within the wage system) drive the development of the *means of production* (tooling machines, steam engines, etc.) and not the other way around, as techno-deterministic readings have been claiming then and now by centring the Industrial Revolution around technological innovation only. Of these principles of labour analysis Marx made also something else: he considered the cooperation of labour not only as a principle to explain the design of machines but also to define the political centrality of what he called the *Gesamtarbeiter*, the *general worker*. The figure of the general worker was a way of acknowledging the machinic dimension of living labour and confronting the 'vast automaton' of the industrial factory on the same scale of complexity. Eventually, it was also a necessary figure to ground, on a more solid politics, the ambivalent idea of the *general intellect* that Ricardian socialists such as William Thompson and Thomas Hodgskin pursued, as seen in chapter 4.

David Noble, *Forces of Production: A Social History of Industrial Automation*, New York: Oxford University Press, 1984.

From the assembly lines to pattern recognition

This book has provided an expanding history of the division of labour and its metrics as a way to identify the operative principle of AI in the long run. As we have seen, at the turn of the nineteenth century, the more the division of labour extended into a globalised world, the more troublesome its management became, requiring new techniques of communication, control, and 'intelligence'. While, within the manufactory, labour management could be still sketched in a simple flow chart and measured by a clock, it was highly complicated to visualise and quantify what Émile Durkheim, already in 1893, defined as 'the division of social labour'.[5] The 'intelligence' of the factory's master could no longer survey the entire production process in a single glance; now, only the infrastructures of communication could achieve this role of supervision and quantification. New mass media, such as the telegraph, telephone, radio, and television networks made possible communication across countries and continents, but they also opened up new perspectives on society and collective behaviours. As seen in chapter 5, James Beniger aptly described the rise of information technologies as a 'control revolution' that proved necessary in that period for governing the economic boom and commercial surplus of the Global North. After World War II, the control of this extended logistics became the concern of a new discipline of the military that bridged mathematics and management: operations research. However, it should be considered that also the transformations of the working class within and across countries, marked by cycles of urban conflicts and decolonial struggles, were among the factors that prompted the rise of these new technologies of control. Chapter 6 endeavoured to trace the historical coincidence between cybernetic projects of self-organisation and the social drives to self-organisation after World War II, as exemplified by countercultural and anti-authoritarian movements.

The scale shift of labour composition from the nineteenth to the twentieth centuries affected also the *logic of automation*, that is, the scientific paradigms involved in this transformation. The relatively simple industrial division of labour and its seemingly rectilinear

---

5 See Émile Durkheim, *De la division du travail social*, Paris: Félix Alcan, 1893, translated as *The Division of Labor in Society*, New York: Free Press, 1984.

assembly lines could easily be compared to a simple *algorithm*, a rule-based procedure with an 'if/then' structure which has its equivalent in the logical form of *deduction*. Deduction, not by coincidence, is the logical form that via Leibniz, Babbage, Shannon, and Turing innervated into electromechanical computation and eventually symbolic AI. Deductive logic is useful for modelling simple processes, but not systems with a multitude of autonomous agents, such as society, the market, or the brain. In these cases, deductive logic is inadequate because it would explode any procedure, machine, or algorithm into an exponential number of instructions (see chapter 7). Out of similar concerns, cybernetics started to investigate self-organisation in living beings and machines to simulate order into high-complexity systems that could not be easily organised according to hierarchical and centralised methods. This was fundamentally the rationale behind connectionism and artificial neural networks (as discussed in chapter 9) and also early research on distributed networks of communication such as Arpanet (the progenitor of the internet).

Across the twentieth century, many other disciplines recorded the growing complexity of social relations. The twin concepts of *Gestalt* and *pattern*, for instance, as employed respectively by Kurt Lewin and Friedrich Hayek and described earlier in this book, were an example of how psychology and economics responded to a new composition of society. Lewin introduced holistic notions such as *force field* and *hodological space* to map group dynamics at different scales between the individual and the mass society.[6] Meanwhile, as we saw in chapter 8, Hayek hijacked the disparate notion of pattern in order to sketch a theory of the market and the mind based on radical individualism.

French thought has been particularly fertile and progressive in this direction. The philosophers Gaston Bachelard and Henri Lefebvre proposed, for example, the method of *rhythmanalysis*, as a study of social rhythms in the urban space (which Lefebvre described according to the four typologies of arrhythmia, polyrhythmia, eurhythmia, and

---

6 See Kurt Lewin, 'Die Sozialisierung des Taylorsystems: Eine grundsätzliche Untersuchung zur Arbeits- und Berufspsychologie', *Schriftenreihe Praktischer Sozialismus*, vol. 4, Berlin: 1920. See also Simon Schaupp, 'Taylorismus oder Kybernetik? Eine kurze ideengeschichte der algorithmischen arbeitssteuerung', *WSI-Mitteilungen* 73, no. 3, (2020): 201–8.

isorhythmia).⁷ In a similar way, French archaeology engaged with the study of expanded forms of social behaviour in ancient civilisations. For instance, the paleoanthropologist André Leroi-Gourhan, together with others, introduced the idea of the operational chain (*chaîne opératoire*) to explain the way pre-historic humans produced utensils.⁸ At the culmination of this long tradition of *diagrammatisation* of social behaviours in French thought, Gilles Deleuze wrote his famous 'Postscript on the Society of Control', which declared that power was no longer concerned with the discipline of individuals but with the control of *dividuals*, that is of the fragments of an extended and deconstructed body.⁹

Lewin's force fields, Lefebvre's urban rhythms, and Deleuze's dividuals can be seen as predictions of the principles of *algorithmic governance* which have been established with the network society and its vast data centres since the late 1990s. The 1998 launch of Google's PageRank algorithm – a method for organising and searching the chaotic hypertext of the web – is considered, by convention, the first large-scale elaboration of 'big data' from digital networks.¹⁰ These techniques for network mapping have become nowadays ubiquitous: Facebook, for instance, uses the Open Graph protocol to quantify the networks of human relations that feed the attention economy of its platform.¹¹ The US military

---

7 Gaston Bachelard, *La dialectique de la durée*, Paris: Boivin & Cie, 1936, translated as *The Dialectic of Duration*, New York: Rowman & Littlefield, 2016; Henri Lefebvre, *Éléments de rythmanalyse*, Paris: Éditions Syllepse, 1992, translated as *Rhythmanalysis: Space, Time and Everyday Life*, London: Continuum, 2004. Bachelard took the term 'rhythmanalysis' from the Portuguese philosopher Lucio Alberto Pinheiro dos Santos.

8 See Frederic Sellet, 'Chaîne opératoire: The Concept and Its Applications', *Lithic Technology* 18, nos. 1–2 (1993): 106–12.

9 Gilles Deleuze, 'Postscript on the Society of Control', *October* 59 (1992); David Savat, 'Deleuze's Objectile: From Discipline to Modulation', in *Deleuze and New Technology*, ed. Mark Poster and David Savat, Edinburgh: Edinburgh University Press, 2009.

10 Matthew L. Jones, 'Querying the Archive: Data Mining from Apriori to PageRank', in *Science in the Archives*, ed. Lorraine Daston, Chicago: University of Chicago Press, 2017; Matteo Pasquinelli, 'Google's PageRank Algorithm: A Diagram of Cognitive Capitalism and the Rentier of the Common Intellect', in *Deep Search*, ed. Konrad Becker and Felix Stalder, London: Transaction Publishers: 2009.

11 Irina Kaldrack and Theo Röhle, 'Divide and Share: Taxonomies, Orders, and Masses in Facebook's Open Graph', *Computational Culture* 4 (November 2014); Tiziana Terranova, 'Securing the Social: Foucault and Social Networks', in *Foucault and the History of Our Present*, ed. S. Fuggle, Y. Lanci, and M. Tazzioli, London: Palgrave Macmillan, 2015.

has been using its own controversial techniques of *pattern-of-life analysis* to map social networks in war zones and to identify targets of drone strikes which, as known, have killed innocent civilians.[12] More recently, gig economy platforms and logistics giants such as Uber, Deliveroo, Wolt, and Amazon started to trace their fleet of riders and drivers via geolocation apps.[13] All these techniques are part of the new field of 'people analytics' (also known as 'social physics' or 'psychographics'), which is but the application of statistics, data analytics, and machine learning to the problem of labour power in post-industrial society.[14]

The automation of psychometrics, or general intelligence

If attractive concepts such as 'pattern', 'Gestalt', or 'model' are not situated in an economic perspective (an opportunity Hayek, for instance, did not miss), their use may easily turn into a self-referential culturalist exercise. The division of labour as much as the design of machines and algorithms are not abstract forms per se but means for measuring labour and social behaviours and discriminating people according to their productive capacity. As the Babbage principles outlined in chapter 2 indicate, any division of labour entails a metrics: a measurement of workers' performativity and efficiency, but also a judgement about classes of skill, which involves an implicit social hierarchy. Metrics of labour were introduced to assess what is and is not productive, to manipulate a social asymmetry while declaring an equivalence to the money system. During the modern age, factories, barracks, and

---

12 Grégoire Chamayou, *A Theory of the Drone*, New York: New Press, 2014, ch. 5: 'Pattern-of-Life Analysis'. See also Matteo Pasquinelli, 'Metadata Society', keyword entry in *Posthuman Glossary*, ed. Rosi Braidotti and Maria Hlavajova, London: Bloomsbury, 2018; Matteo Pasquinelli, 'Arcana Mathematica Imperii: The Evolution of Western Computational Norms', in *Former West*, ed. Maria Hlavajova and Simon Sheikh, Cambridge, MA: MIT Press, 2017.

13 Andrea Brighenti and Andrea Pavoni, 'On Urban Trajectology: Algorithmic Mobilities and Atmocultural Navigation', *Distinktion: Journal of Social Theory* 24, no. 1 (2001): 40–63.

14 L. M. Giermindl et al., 'The Dark Sides of People Analytics: Reviewing the Perils for Organisations and Employees', *European Journal of Information Systems* 33, no. 3 (2022): 410–35. See also Alex Pentland, *Social Physics: How Social Networks Can Make Us Smarter*, London: Penguin, 2015.

hospitals have pursued a discipline and organisation of bodies and minds with similar methods, as Michel Foucault sensed among others.

At the end of the nineteenth century, the metrology of labour and behaviours found an ally in a new field of statistics: psychometrics. Psychometrics had the purpose of measuring the skills of the population in resolving basic tasks, making statistical comparisons on cognitive tests rather than taking measurements of physical performance as in the earlier field of psychophysics.[15] As part of the controversial legacy of Alfred Binet, Charles Spearman, and Louis Thurstone, psychometrics can be considered one of the main genealogies of statistics, which has never been a neutral discipline so much as one concerned with the 'measure of man', the institutions of norms of behaviour, and the repression of abnormalities.[16] The transformation of the metrics of labour into the *psychometrics of labour* is a key passage for both management and technological development in the twentieth century. It is telling, as we saw in chapter 9, that in designing the first artificial neural network perceptron Frank Rosenblatt was not only inspired by theories of neuroplasticity but also by tools of multivariable analysis that psychometrics imported into US psychology in the 1950s.

From this perspective, this book attempted to clarify how the project of AI has actually emerged from the automation of the psychometrics of labour and social behaviours rather than the quest to solve the 'enigma' of intelligence. In a concise summary of the history of AI, one could say that the mechanisation of the 'general intellect' of the industrial age into the 'artificial intelligence' of the twenty-first century was made possible thanks to the statistical measurement of skill, such as Spearman's 'general intelligence' factor and its subsequent automation into artificial neural networks.

---

15 The word 'statistics' originally means the knowledge possessed by the state about its own affairs and territories: a knowledge that had to be kept secret. Michel Foucault, *Security, Territory, Population: Lectures at the Collège de France 1977–1978*, trans. Graham Burchell, London: Palgrave Macmillan, 2009, 274.

16 For the influence of brain metrology on the history of AI, see Simon Schaffer: 'Judgements that machines are intelligent have involved techniques for measuring brains' outputs. These techniques show how discretionary behaviour is connected with status of those who rely on intelligence for their social legitimacy.' Simon Schaffer, 'OK Computer', in *Ansichten der Wissenschaftsgeschichte*, ed. Michael Hagner, Frankfurt: Fischer, 2001, 393–429. On early metrology of the nervous system, see Henning Schmidgen, *The Helmholtz Curves: Tracing Lost Times*, New York: Fordham University Press, 2014.

If in the industrial age the machine was considered as an embodiment of science, knowledge, and the 'general intellect' of workers, in the information age artificial neural networks became the first machine to encode 'general intelligence' into statistical tools – at the beginning, specifically, to automate pattern recognition as one of the key tasks of 'artificial intelligence'. In short, the current form of AI, machine learning, is the automation of the statistical metrics which were originally introduced to quantify cognitive, social, and work-related abilities. The application of psychometrics through information technologies is not a phenomenon unique to machine learning. The 2018 Facebook–Cambridge Analytica data scandal, in which the consulting firm was enabled to collect the personal data of millions without their consent, is a reminder of how large-scale psychometrics is still used by corporate and state actors in the attempt to predict and manipulate collective behaviours.[17]

Given their legacy in the statistical tools of nineteenth-century biometrics, it is also not surprising that deep artificial neural networks have recently unfolded into advanced techniques of surveillance, such as facial recognition and pattern-of-life analysis. Critical AI scholars such as Ruha Benjamin and Wendy Chun, among others, have exposed the racist origins of these techniques of identification and profiling that, like psychometrics, almost represent technical proof of the social bias of AI.[18] They have rightly identified the power of discrimination at the core of machine learning, and how this aligns it with the apparatuses of normativity of the modern age, including the questionable taxonomies of medicine, psychiatry, and criminal law.[19]

The metrology of intelligence pioneered in the late nineteenth century, with its implicit and explicit agenda of social and racial segregation, still operates at the core of AI to discipline labour and replicate productive hierarchies of knowledge. The rationale of AI is therefore not only the automation of labour but the reinforcement of these social hierarchies in

---

17 Luke Stark, 'Algorithmic Psychometrics and the Scalable Subject', *Social Studies of Science* 48, no. 2 (2018): 204–31.

18 Ruha Benjamin, *Race after Technology: Abolitionist Tools for the New Jim Code*, Cambridge: Polity, 2019; Wendy Chun, *Discriminating Data: Correlation, Neighborhoods, and the New Politics of Recognition*, Cambridge, MA: MIT Press, 2021.

19 For the imbrication of colonialism, racism, and digital technologies, see Jonathan Beller, *The World Computer: Derivative Conditions of Racial Capitalism*, Durham, NC: Duke University Press, 2021; Seb Franklin, *The Digitally Disposed: Racial Capitalism and the Informatics of Value*, Minneapolis: University of Minnesota Press, 2021.

an indirect way. By implicitly declaring what can be automated and what cannot, AI has imposed a new metrics of intelligence at each stage of its development. But to compare human and machine intelligence implies also a judgement about which human behaviour or social group is more intelligent than another, which workers can be replaced and which cannot. Ultimately, AI is not only a tool for automating labour but also for imposing standards of *mechanical intelligence* that propagate, more or less invisibly, social hierarchies of knowledge and skill. As with any previous form of automation, AI does not simply replace workers but displaces and restructures them into a new social order.

The automation of automation

Looking carefully at how statistical tools that were conceived to rate cognitive skills and discriminate between people's productivity turned into algorithms, a more profound aspect of automation becomes apparent. In fact, the study of the metrology of labour and behaviours reveals that automation emerges in some cases from the transformation of the measurement instruments themselves into kinetic technologies. Tools for labour quantification and social discrimination have become 'robots' in their own right. Before psychometrics, one could refer to how the clock used to measure labour time in the factory was later implemented by Babbage for the automation of mental labour in the Difference Engine (see chapter 2). Cyberneticians such as Norbert Wiener still considered the clock as a key model for both the brain and the computer. In this respect, the historian of science Henning Schmidgen has noted how the chronometry of nervous stimuli contributed to the consolidation of brain metrology and also McCulloch and Pitts's model of neural networks.[20] The theory of automation which this book has illustrated, then, does not point only to the emergence of machines from the logic of labour management but also from the instruments and metrics for quantifying human life in general and making it productive.

---

20 Henning Schmidgen, 'Cybernetic Times: Norbert Wiener, John Stroud, and the "Brain Clock" Hypothesis', *History of the Human Sciences* 33, no. 1 (2020), 80–108. On cybernetics and the 'measurement of rationality', see Orit Halpern, *Beautiful Data: A History of Vision and Reason since 1945*, Durham, NC: Duke University Press, 2015, 173.

Conclusion: The Automation of General Intelligence 247

This book has sought to show that AI is the culmination of the long evolution of labour automation and quantification of society. The statistical models of machine learning do not appear, in fact, to be radically different but rather homologous to the design of industrial machines: they are indeed constituted by the same analytical intelligence of tasks and collective behaviours, albeit with a higher degree of complexity (i.e., number of parameters). Like industrial machines whose design gradually emerged through routine tasks and trial-and-error adjustments, machine learning algorithms adapt their internal model to the patterns in the training data through a comparable trial-and-error process. The *design* of a machine as well as the *model* of a statistical algorithm can be said to follow a similar logic: both are based on the imitation of an external configuration of space, time, relations, and operations. In the history of AI, this was as true of Rosenblatt's perceptron (which aimed to record the gaze's movements and spatial relations of the visual field) as of any other machine learning algorithm nowadays (e.g., support vector machines, Bayesian networks, transformer models).

Whereas the industrial machine embodies the diagram of the division of labour in a determined way (think of the components and limited 'degrees of freedom' of a textile loom, a lathe, or a mining excavator), machine learning algorithms (especially recent AI models with a vast numbers of parameters) can imitate complex human activities.[21] Although with problematic levels of approximation and bias, a machine learning model is an adaptive artefact that can encode and reproduce the most diverse configurations of tasks. For example, one and the same machine learning model can emulate the movement of robotic arms in assembly lines as much as the driver's operations in a self-driving car; the same model can also translate between languages as much as describe images with colloquial words.

The rise of large foundation models in recent years (e.g., BERT, GPT, CLIP, Codex) demonstrates how one single deep learning algorithm can be trained on one vast integrated dataset (comprising text, images, speech, structured data, and 3-D signals) and used to automate a wide

---

21 In mechanics, the degrees of freedom (DOF) of a system such as a machine, a robot, or a vehicle is the number of independent parameters that define its configuration or state. Bicycles are usually said to have two degrees of freedom. A robotic arm can have many. A large machine learning model such as GPT can feature more than a trillion.

range of so-called downstream tasks (question answering, sentiment analysis, information extraction, text generation, image captioning, image generation, style transfer, object recognition, instruction following, etc.).[22] For the way in which they have been built on large repositories of cultural heritage, collective knowledge, and social data, large foundation models are the closest approximation of the mechanisation of the 'general intellect' which was envisioned in the industrial age. An important aspect of machine learning that foundation models demonstrate is that the automation of individual tasks, the codification of cultural heritage, and the analysis of social behaviours have no technical distinction: they can be performed by the one and same process of statistical modelling.

In conclusion, machine learning can be seen as the project to automate the very process of machine design and model making – which is to say, the automation of the labour theory of automation itself. In this sense, machine learning and, specifically, large foundation models represent a new definition of the Universal Machine, for their capacity is not just to perform computational tasks but to imitate labour and collective behaviours at large. The breakthrough that machine learning has come to represent is therefore not just the 'automation of statistics', as machine learning is sometimes described, but the *automation of automation*, bringing this process to the scale of collective knowledge and cultural heritage.[23] Further, machine learning can be considered as the technical proof of gradual integration of labour automation with social governance. Emerging out of the imitation of the division of labour and psychometrics, machine learning models have gradually evolved towards an integrated paradigm of governance that corporate data analytics and its vast datacentres well exemplify.

At this point of analysis, it is important to mention that intrinsic limits affect the current form of machine learning. In a paper published in September 2021, Neil Thompson and other computer scientists argue

---

22 Rishi Bommasani et al., *On the Opportunities and Risks of Foundation Models*, Center for Research on Foundation Models at the Stanford Institute for Human-Centered Artificial Intelligence, 2021, crfm.stanford.edu.

23 For a different reading of the automation of automation, see Pedro Domingos, *The Master Algorithm: How the Quest for the Ultimate Learning Machine Will Remake Our World*, New York: Basic Books, 2015; Luciana Parisi, 'Critical Computation: Digital Automata and General Artificial Thinking', *Theory, Culture, and Society* 36, no. 2 (March 2019): 89–121.

that the error-correction techniques of deep learning have reached a computational limit and are unable to grow without paying exorbitant costs of energy and hardware resources which not even big corporations could soon afford.[24] The issue of computational explosion, which cyclically reappears in the history of AI, this time affects artificial neural networks. These findings, which can be generalised to other algorithms and error-correction techniques, simultaneously prove that the 'intelligence explosion' of AI is a mirage. As such, when critical theory engages with cartoonish campaigns to discover the 'alien intelligence' hidden in the black box of AI, it often neglects this logical limit. What scholars perceive as 'alien intelligence' is, more prosaically, the game of statistical correlations at a very large scale. There is no evidence of a 'singularity' phenomenon in this game of correlations and, given its computational constraints, current AI runs no risk of becoming the malevolent 'superintelligence' of which Oxford scholar Nick Bostrom has warned.

Undoing the master algorithm

Given the growing size of datasets, the training costs of large models, and the monopoly of the cloud infrastructure that is necessary to host such models by a few companies such as Amazon, Google, and Microsoft (and their Asian counterparts Alibaba and Tencent), it has become evident to everyone that the sovereignty of AI remains a tough affair of geopolitical scale. Moreover, the confluence of different apparatuses of governance (climate science, global logistics, and even health care) towards the same hardware (cloud computing) and software (machine learning) signals an even stronger trend to monopolisation. Aside from the notorious issue of power accumulation, the rise of data monopolies points at a phenomenon of technical convergence that is key to this book: the means of labour have become the same ones of its measurement, and, likewise, the means of management and logistics have become the same ones of economic planning.

This became evident also during the recent COVID-19 pandemic,

---

24 Neil Thompson et al., 'Deep Learning's Diminishing Returns', *IEEE Spectrum*, 24 September 2021, spectrum.ieee.org; Neil Thompson et al., 'The Computational Limits of Deep Learning', 10 July 2020, preprint, arxiv.org.

when a large infrastructure for tracking, measuring, and forecasting social behaviours was established.[25] This infrastructure, unprecedented in the history of health care and biopolitics, however, was not created *ex nihilo* but built upon existing digital platforms that orchestrate most of our social relations. Particularly during the lockdowns, the same digital medium was used for working, shopping, communicating with family and friends, and eventually health care. Digital metrics of the social body such as geolocation and other metadata were key for the predictive models of the global contagion, but they have been long in use for tracking labour, logistics, commerce, and education. Philosophers such as Giorgio Agamben have claimed that this infrastructure extended the state of emergency of the pandemic, while in fact its deployment to health care and biopolitics continues decades of monitoring the economic productivity of the social body which passed unnoticed to many.[26]

The technical convergence of data infrastructures reveals also that contemporary automation is not just about the automation of an individual worker, as in the stereotypical image of the humanoid robot, but about the automation of the factory's masters and managers, as happens in the gig economy platforms. From the giants of logistics (Amazon, Alibaba, DHL, UPS, etc.) and mobility (Uber, Share Now, Foodora, Deliveroo) to social media (Facebook, TikTok, Twitter) – platform capitalism is a form of automation that in reality does not replace workers but multiplies and governs them anew. It is not so much about the automation of labour this time as it is about the automation of management. Under this new form of *algorithmic management*, we are all rendered as *dividual workers* of a vast automaton comprised of global users, 'turkers', carers, drivers, and riders of many sorts. The debate on the fear that AI fully replaces jobs is misguided: in the so-called platform economy, in reality, algorithms replace management and multiply precarious jobs. Although the revenues of the gig economy remain minoritarian in relation to traditional local sectors, by using the same infrastructure worldwide these platforms have established monopoly

---

25 See Breaking Models: Data Governance and New Metrics of Knowledge in the Time of the Pandemic, workshop, Max Planck Institute for the History of Science, Berlin, and KIM research group, University of Arts and Design, Karlsruhe, 24 September 2021, kim.hfg-karlsruhe.de/breaking-models.

26 Giorgio Agamben, *Where Are We Now? The Epidemic as Politics*, trans. V. Dani, Lanham, MA: Rowman & Littlefield, 2021.

Conclusion: The Automation of General Intelligence 251

positions. In conclusion, the power of the new 'master' is not about the automation of individual tasks but the management of the social division of labour. Against Alan Turing's prediction, it was the master, not the worker, that the robot came to replace first.[27]

One wonders what the chance of political intervention in such technologically integrated space would be, and whether the call to 'redesign AI' that grassroots and institutional initiatives advocate for is either reasonable or practicable. This call should first respond to the more pressing question: How is it possible to 'redesign' large-scale monopolies of data and knowledge?[28] As big companies such as Amazon, Walmart, and Google have conquered a unique access to the needs and problems of the whole social body, a growing movement is asking not just to make these infrastructures more transparent and accountable but actually to collectivise them as public services (as Fredric Jameson has suggested, among others), or having them replaced by public alternatives (as Nick Srnicek has advocated).[29] But what would be a different way to design such alternatives?

As this book's theory of automation has suggested, any technology and institutional apparatus, including AI, is a crystallisation of a productive social process. Problems arise because such crystallisation 'ossifies' and reiterates past structures, hierarchies, and inequalities. To criticise and deconstruct complex artefacts such as AI monopolies, first we should engage in a meticulous work of *deconnectionism*, undoing – step by step, file by file, dataset by dataset, piece of metadata by piece of metadata, correlation by correlation, pattern by pattern – the social and economic fabric that constitutes them in origin. This work is already

---

27 See Min Kyung Lee et al., 'Working with Machines: The Impact of Algorithmic and Data-Driven Management on Human Workers', in *Proceedings of the 33rd Annual ACM Conference on Human Factors in Computing Systems*, Association for Computing Machinery, New York, 2015; Sarah O'Connor, 'When Your Boss Is an Algorithm', *Financial Times*, 8 September 2016. See also Alex Wood, 'Algorithmic Management Consequences for Work Organisation and Working Conditions', No. 2021/07, European Commission JRC Technical Report, Working Papers Series on Labour, Education, and Technology, 2021.

28 See Daron Acemoglu (ed.), *Redesigning AI*, Cambridge, MA: MIT Press, 2021.

29 Leigh Phillips and Michal Rozworski, *The People's Republic of Walmart: How the World's Biggest Corporations Are Laying the Foundation for Socialism*, London: Verso, 2019. See also Frederic Jameson, *Archaeologies of the Future: The Desire Called Utopia and Other Science Fictions*, London: Verso, 2005, 153n22; Nick Srnicek, *Platform Capitalism*, Cambridge: Polity Press, 2017, 128.

being advanced by a new generation of scholars who are dissecting the global production pipeline of AI, especially those who use methods of *action research*. Notable, among many others, are Lilly Irani's Turkopticon platform, used for 'interrupting worker invisibility' in the gig platform Amazon Mechanical Turk; Adam Harvey's investigation of training datasets for face recognition, which exposed the massive privacy infringements of AI corporations and academic research; or the work of the Politically Mathematics collective from India, who analysed the economic impact of COVID-19 predictive models on the poorest population and reclaimed mathematics as a space of political struggle (see their manifesto quoted at the beginning of this conclusion).[30]

The labour theory of automation is an analytical principle for studying also the new 'eye of the master' which AI monopolies incarnate. However, precisely because of the emphasis on the labour process and social relations that constitute technical systems, it is also a synthetic and 'sociogenic' principle (to use Frantz Fanon and Sylvia Wynter's programmatic term).[31] What is at the core of the labour theory of automation is, ultimately, a *practice of social autonomy*. Technologies can be judged, contested, reappropriated, and reinvented only by moving into the matrix of the social relations that originally constituted them. Alternative technologies should be *situated* in these social relations, in a way not dissimilar to what cooperative movements have done in the past centuries. But building alternative algorithms does not mean to make them more ethical. For instance, the proposal to hard-code ethical rules into AI and robots appears highly insufficient and incomplete because it does not directly address the broad political function of automation at their core.[32]

What is needed is neither techno-solutionism nor techno-pauperism, but instead a *culture of invention*, design and planning which cares for communities and the collective, and never entirely relinquishes agency

---

30 See the websites turkopticon.net, exposing.ai, and politicallymath.in.

31 Sylvia Wynter, 'Towards the Sociogenic Principle: Fanon, Identity, the Puzzle of Conscious Experience, and What It Is Like to Be "Black" ', *National Identities and Sociopolitical Changes in Latin America* (2001): 30–66. See also Luciana Parisi, 'Interactive Computation and Artificial Epistemologies', *Theory, Culture, and Society* 38, nos. 7–8 (October 2021): 33–53.

32 See Frank Pasquale, *New Laws of Robotics: Defending Human Expertise in the Age of AI*, Cambridge, MA: Harvard University Press, 2020; Dan McQuillan, 'People's Councils for Ethical Machine Learning', *Social Media+ Society* 4, no. 2 (2018).

and intelligence to automation. The first step of technopolitics is not technological but political. It is about emancipating and decolonising, when not abolishing as a whole, the organisation of labour and social relations on which complex technical systems, industrial robots, and social algorithms are based – specifically their inbuilt wage system, property rights, and identity politics. New technologies for labour and society can only be based on this political transformation. It is clear that this process unfolds also by developing not only technical but also political knowledge. One of the problematic effects of AI on society is its epistemic influence – the way in which it renders intelligence as machine intelligence and implicitly fosters knowledge as procedural knowledge. The project of a political epistemology to transcend AI, however, will have to transmute the historical forms of abstract thinking (mathematical, mechanical, algorithmic, and statistical) and integrate them as part of the toolbox of critical thinking itself. In confronting the epistemology of AI and its regime of knowledge extractivism, a different technical mentality, a collective 'counter-intelligence', has to be learned.

Acknowledgements

This book is indebted to a 'collective philosopher' who contributed to its composition along multidimensional trajectories. McKenzie Wark, at the New School in New York, suggested many years ago that I initiate this project for Verso Books. Henning Schmidgen at Weimar University helped me to see, in a truly interdisciplinary way, the complex horizon of thought spanning science and technology studies, historical epistemology, French philosophy, and media studies. In the last months of its redaction, Sascha Freyberg at the Max Planck Institute for the History of Science in Berlin generously intervened to consolidate the manuscript. Special thanks go to Sam Smith, copy editor, and to George MacBeth and Patrick Riechert at the University of Arts and Design Karlsruhe, who patiently revised my non-native English and improved my non-linear way of writing.

In the community of the historians of science and technology, I would like to thank in particular: Pietro Daniel Omodeo, Giulia Rispoli, Matthias Schemmel, Charles Wolfe, and Senthil Babu, including the Politically Mathematics collective and Verum Factum editorial collective. I thank especially students, colleagues, and alumnae of the University of Arts and Design Karlsruhe, the PhD colloquium, and the KIM research group on artificial intelligence and media philosophy: Paolo Caffoni, Céline Condorelli, Alan Diaz, Ariana Dongus, Max Grünberg, Sami Khatib, Maya Indira Ganesh, Arif Kornweitz, James Langdon, Angela Melitopoulos, Simone

Murru, Mariana Silva, Ana Teixeira Pinto, and Füsun Türetken – with a special mention to Bárbara Acevedo Strange, Sebastian Breu, Guillermo Collado Wilkins, Alex Estorick, Yannick Fritz, Vincent Herrmann, Nikos Patelis, Simon Knebl, Jason King, Lukas Rehm, Marco Schröder, Ben Seymour, and Florian Walzel for their technical, graphical, and editorial assistance.

I'm indebted to an international community of AI and media scholars with whom, in the past decade, I have been conspiring on numerous projects: Jon Beller, Claude Draude, Orit Halpern, Adam Harvey, Leonardo Impett, Daniel Irrgang, Vladan Joler, Goda Klumbyte, Oleksiy Kuchansky, Noura Al Moubayed, Fabian Offert, Ranjodh Singh Dhaliwal, Susan Schuppli, Elena Vogman, and Ben Tarnoff. And for their friendship and support: Marie-Luise Angerer, Alan Blackwell, Andreas Broeckmann, Johannes Bruder, Mercedes Bunz, Geoff Cox, Sean Cubitt, Jan Distelmeyer, Daphne Dragona, Paul Feigelfeld, Kristoffer Gansing, Francis Hunger, Joseph Lemelin, Dan McQuillan, Frieder Nake, Luciana Parisi, Jens Schröter, and Andreas Sudmann – and, in particular, Nina Franz, Rebekka Ladewig, Wolfgang Lefèvre, George Nagy, Simon Schaffer, Max Stadler, Siegfried Zielinski, and Paolo Zellini.

I thank especially the organisers of the 2020–21 Cambridge summer school 'Histories of AI: A Genealogy of Power' – Sarah Dillon, Richard Staley, Syed Mustafa Ali, Stephanie Dick, Matthew L. Jones, and Jonnie Penn – for the opportunity to present and discuss my ideas.

The 'political intelligence' upon which this book is based proceeds from a long-standing international collective of friends and comrades: Stefano Harney, Sandro Mezzadra, Toni Negri, Brett Neilson, Nikos Papastergiadis, Judith Revel, Tania Rispoli, Paolo Virno, Jamie Woodcock, and Giovanna Zapperi, among many others. For their support through all these years, I thank Marco Baravalle, Susana Caló, Manuel Disegni, Godofredo Pereira, and Gerard Hanlon.

I thank *e-flux journal* (Julieta Aranda, Amal Issa, Brian Kuan Wood, Anton Vidokle) and the House of World Cultures in Berlin (Anselm Franke, Bern Scherer) for their hospitality across several projects. And, among the many friends of the art community: Elena Agudio, Dorothee Albrecht, Marwa Arsanios, Defne Ayas, Sean Dockray, Gean Moreno, Fiona Geuss, Natasha Ginwala, Nathan Gray, Christoph Gurk, Tom Holert, Iman Issa, Sven Lütticken, Antonia Majaca, Agnieszka Polska,

Johannes Paul Raether, Natascha Sadr Haghighian, Caleb Waldorf, Clemens von Wedemeyer, and Khadija von Zinnenburg Carroll.

This book acknowledges also the fellows, colleagues, and employees of the Staatsbibliothek of Berlin, with whom I invisibly crossed paths for many years, upholding our vow of silence. I thank especially the staff of the library for their support during the COVID-19 pandemic. This book is sincerely addressed to all the ghost workers, bikers, drivers, and coders of the gig economy on the front line of the algorithmic division of labour.

The research of this book was initiated well before my son Giotto was born and finished long after. It was made materially and maternally possible also thanks to the love and care of Wietske on the battlefield of our domestic labour. For Giotto, Wietske, and friends, I venture into the engine rooms of these technological abstractions hoping to spare everyone the need of similar alienating adventures in the future. This book is dedicated to my mother, Anna, who had the strange idea to teach me writing and counting before I started school. It was further exercises in that working-class 'kitchen maths' that, I believe, four decades later, led me to study the progress of these vast social algorithms.

Index

Abraham, Tara, 165
abstract energy, 123
abstract time, 125, 203
abstraction, 13, 24, 30–2, 34–40, 47, 67–8, 91, 109, 120, 169, 172, 174, 181, 190, 194, 201–4, 215–16, 221, 256
   of labour, 121, 125. *See also* labour
   material, 16, 40, 238
   real, 201–2
   reflective, 35–6
Adam, Alison, 10
Agamben, Giorgio, 250
Agar, Jon, 11
Agnicayana, 16, 23–6
Agre, Philip, 10
Alexandre de Villedieu, 41
AlexNet (convolutional neural network), 14, 216, 231
algorithm, 2–3, 6, 10, 12–13, 16, 20, 22–31, 33, 36–7, 40–8, 53, 55–8, 60, 66–70, 73, 94, 104, 130, 134–5, 143, 150–1, 155, 163, 165–6, 168, 198–9, 201, 213–15, 217, 219–20, 222–3, 231–8, 241–3, 245–7, 249–53, 256
   algorismus, 40–1, 48
   classifier, 183, 191, 234
   definition of, 16, 26, 29, 47
   master, 9, 248
   labour as, 16
   social, 4, 28, 238, 253, 256
   training, 213–15, 223, 231
algorithmic culture, 26
algorithmic management, 250–251
algorithmic thinking, 13, 16, 23, 66, 70, 253
alienation, 17, 18, 86, 100, 105, 108–9
al-Khwarizmi, Muhammad ibn Musa, 41
Alquati, Romano, 13–14, 119, 121, 126–9
Altenried, Moritz, 6
Althusser, Louis, 102
Analytical Engine, 17, 45, 54, 58, 67–73, 76
Anaximander, 202
Anaximenes, 202
artificial intelligence (AI), 1–3, 6–15, 17–22, 27, 47, 52–3, 68, 72, 76, 79, 93–4, 100, 130, 134, 136–7, 146, 151–2, 154–6, 158–60, 164, 166, 169, 175–6, 183, 196, 198, 201, 206, 217–20, 222, 224–5, 227, 229–30, 232, 234–6, 238, 240–1, 244–7, 249–53
   history of, 8–14
Asaro, Peter, 143
Ash, Mitchell G., 162
Ashby, Ross, 140–1, 206, 221

Ashley, Mike, 97
Ashworth, William J., 76
Aspromourgos, Tony, 62
Atanasoski, Neda, 8, 229
Automatic Computing Engine (ACE), 7
automation, 2, 8–9, 14–17, 20–2, 45, 48, 53, 58, 64, 68, 70, 94, 96, 102, 122, 129, 135, 141, 143, 147, 154, 156–60, 162–4, 199, 207, 225–7, 229–30, 233-6
  labour theory of, 6, 11, 22, 239–40, 243, 245–8, 250–3. *See also under* labour theory
  of management, 8, 149, 236, 250–1
  of mental labour, 16–17, 139, 154, 197, 233
  of visual labour, 17

Babbage, Charles, 4–6, 9, 14, 17–18, 45, 51–69, 71–6, 81, 83, 88, 94, 98, 103–6, 109, 114, 116, 118–20, 139, 153, 191, 197–8, 225, 239, 241, 243, 246
Babbage, Henry P., 66
Babbage principle, 14, 34
Babbage's mechanical notation, 66–8, 74
Babu, Senthil, 27
Bachelard, Gaston 13, 241–2
Bajohr, Hannes, 164
Ballard, J. G., 134
Barnes, Barry, 82
Bates, David, 176, 179, 187, 196
Baum, Joan, 68
Beamish, Rob, 18
Beaujean, Alexander, 228
Beer, Stafford, 149
Bell, Bill, 97
Beller, Jonathan, 5, 233, 245
Benanav, Aaron, 21
Beniger, James, 124, 126, 159, 240
Benjamin, Ruha, 10, 245
Benson, Nicholas, 228
Berg, Maxine, 79–82, 87, 89, 114
Bernoulli, Daniel, 45, 69
BERT (language model), 247
Big Data, 10, 12, 55, 199, 235, 242
Binet, Alfred, 224, 244
Black, Edwin, 53

Block, H. D., 216
Blum, Harold Francis, 92
Boden, Margaret, 10, 19, 134, 146
Bommasani, Rishi, 248
Boncompagni, Baldassarre, 41
Boolean logic, 46, 136, 166
Bostrom, Nick, 9, 249
Bowden, Bertram, 68
Brake, Laurel, 97
Brautigan, Richard, 157
Braverman, Harry, 5, 13–14, 119–20
Brighenti, Andrea, 243
Brougham, Henry, 81
Brown, Elspeth, 5
Brown, Tom, 232
Bruderer, Herbert, 220
Bücher, Karl, 32
Büchner, Georg, 99
Buxton, Harry Wilmot, 55
Byron, George Gordon (Lord Byron), 68–9

Caffentzis, George, 18
calculus, differential, 26–7, 43, 69, 164, 194, 215
Canguilhem, Georges, 13, 179, 187
Cannon, Walter Bradford, 141
Cardon, Dominique, 14
Carlyle, Thomas, 81
Cassirer, Ernst, 32
Castelle, Michael, 137, 235
Chabert, Jean-Luc, 16, 26, 30–1, 43, 47
chaîne opératoire (operational chain), 25, 242
Chamayou, Grégoire, 243
Chollet, François, 164
Chronocyclegraph, 5
Chun, Wendy Hui Kyong, 10, 245
Clark, William, 9, 214
Clarke, Bruce, 13
classification, 14, 20, 117, 162–3, 166, 168, 183, 191–2, 194–8, 204, 207–10, 212, 216, 222–5, 230, 234
CLIP (image-language model), 247
Codex (language model), 247
Cohen, Hermann, 28
Cointet, Jean-Phillipe, 14
Colebrooke, Henry, 52

Index

computing network, 51, 147, 150, 168, 184, 209, 213, 222
connectionism, 15, 17, 19–20, 134, 141, 149, 151, 169, 182–6, 191, 196, 198, 200, 202, 241
  deconnectionism, 251
  Gestalt, 196
  logical, 197
  mercantile, 185
  statistical, 197
Copeland, B. Jack, 7, 20, 140
Crossley, John N., 41
cultural techniques (*Kulturtechniken*), 31, 103
cybernetics, 10–11, 14, 17, 19, 74, 121–2, 126–9, 135–7, 139–40, 147, 149, 152–60, 162, 164, 168, 171–3, 176, 179, 183–4, 187, 194, 196, 221, 227–8, 234, 241, 246
Cyclogram, 5

D'Alembert, Jean le Rond, 43
Damerow, Peter, 13, 24, 34–40, 43, 48, 89, 91, 120, 202
Daston, Lorraine, 3, 4, 17, 24, 55, 57, 59, 68, 108, 242
de Prony, Gaspard, 4, 45, 53, 55–7, 60
De Vecchi, Nicolò, 193
deep learning, 14–15, 20, 134, 137, 150, 162, 164–5, 169, 178, 215–17, 219, 231, 235, 247, 249
Defence Advanced Research Projects Agency (DARPA), 26
Deleuze, Gilles, 32, 102, 242
Descartes, René, 45
Desrosières, Alain, 228
Di Iorio, Francesco, 186
diaschisis, 187
Dick, Stephanie, 219
Difference Engine, 14
Domingos, Pedro, 9, 248
Dotzler, Bernhard, 18
Dreyfus, Stuart, 14–15, 152
Dreyfus, Hubert, 10, 14–15, 152
Du Bois-Reymond, Emil, 133, 137
Dupuy, Jean-Pierre, 196
Durkheim, Émile, 6, 240
Dyer-Witheford, Nick, 101

Eddington, Arthur Stanley, 92
Edwards, Paul, 10
Electronic Profile Analyzing Computer (EPAC), 225
Engels, Friedrich, 4, 77–8, 99
Ensmenger, Nathan, 29
epistemic colonialism, 27
epistemology, 11, 15, 181, 182, 195, 198, 202
  cybernetic, 154
  experimental, 218
  feminist, 13
  genetic, 34, 38,
  historical, 12–13, 34, 39, 84, 92, 154
  political, 13, 39, 201, 253
Erickson, Paul, 11, 146, 185
Essinger, James, 69

factory, 5, 8, 52, 58, 63–4, 74, 76, 78, 85, 107, 110, 117, 157, 191, 239, 246
  automated, 128
  computer, 14, 126–7
  cybernetic, 149
  digital, 17
  social, 128
Fanon, Frantz, 252
Farley, B. G., 147, 214
Federici, Silvia, 13
Fein, Louis, 199, 217
Feldhay, Rivka, 66
Ferguson, Adam, 111
Fibonacci, 41
Finkelstein, David, 97
Fisher, Roland, 226, 229
Forsyth, Richard, 231
Forsyth, George, 155
Foucault, Michel, 4, 179, 187, 242, 244
Francis, Jo, 71,
Franklin, Seb, 245
Fuegi, John, 71
Fukushima, Kunihiko, 216

Gaia hypothesis, 138
Galilei, Galileo, 90,
Galison, Peter, 92, 175
Galloway, Alex, 144
Galton, Francis, 229

Garvey, Shunryu, 146
Gastev, Aleksei, 5
general intellect, 5, 18–19, 83, 85, 88, 95–6, 98–105, 110, 112, 116–19, 190, 239, 244–5, 248
General Problem Solver, 31
general purpose computer, 70
general worker (*Gesamtarbeiter*), 19, 88, 94, 99, 116–17, 121, 191, 239–40
Geoghegan, Bernard, 9, 129
Gerovitch, Slava, 185
Gestalt perception, 38
  isomorphism in, 169, 171–2
  Prägnanz principle in, 163–4
ghost workers, 8, 21, 256
Gibson, William, 182
Giedion, Sigfried, 5
Giermindl, L. M., 243
Gigerenzer, Gerd, 226–7, 229
Gilbreth, Frank, 5
Gilbreth, Lillian, 5
Gödel, Kurt, 46
Goethe, Johan Wolfgang von, 193
Goldstein, Daniel, 229,
Goldstein, Kurt, 172, 179, 184, 187–90, 221–2
Golinski, Jan, 9
GoogleNet (convolutional neural network), 231
Gould, Stephen Jay, 21, 229
GPT (Generative Pre-trained Transformer), 213, 219, 232, 247
Grab, Walter, 99
Graeber, David, 21
Gramsci, Antonio, 1, 12
Grattan-Guinness, Ivor, 56, 68
Gray, Mary, 8
Grossmann, Henryk, 13, 45, 92, 107
Guattari, Felix, 32, 102

Hadden, Richard, 11
Haigh, Thomas, 165
Halpern, Orit, 246
Hancher, Michael, 97
Harding, Sandra, 13
Hardt, Michael, 101, 204
Hartley, Ralph V., 129
Harvey, Adam, 252
Haug, Wolfgang Fritz, 103

Hawkins, Andrew J., 3
Hay, John, 212
Hayek, Friedrich, 20, 148–9, 159, 181–204, 219, 221, 241, 243
Heath, William, 96–7
Hebb, Donald, 141–3, 150, 183, 206, 212, 221
Hebbian rule, 142, 143, 212, 206
Hegel, Georg Wilhelm Friedrich, 36–8, 99, 121, 124–5, 202–3, 239
Heims, Steve, 136, 162, 164, 168, 171
Heinrich, Michael, 101, 118, 163
Hellebust, Rolf, 5
Helmholtz, Hermann von, 137–8, 244
Henry, Alan S., 41
Herschel, John, 75–6
Herzog, Don, 97
Hessen, Boris, 13, 45, 92, 107
Hinton, Geoffrey, 14
history of science and technology, 12, 27, 92, 106
Hodgskin, Thomas, 18, 75, 84, 86–8, 93, 98–9, 108–9, 113–15, 119, 239
Hof, Ebbe C., 187
Hoffmann, Christoph, 137
Hollerith machine, 11
homeostat, 140–1, 206
Hubel, David, 175, 216
hyperplane, 222–3

IBM 704, 210–11
ImageNet (training dataset), 14, 134
information, 3, 11, 19, 29, 46–7, 53–4, 100, 120, 121–30, 139–40, 148, 154–5, 184, 198
  self-organising, 134–7, 150–2
intelligence
  general, 21, 228, 232, 237, 243–5
  quotient of (IQ), 21, 228
  statistical, 223
  superintelligence, 9, 159, 249
Irani, Lilly, 8, 252

Jackson, Hughlings, 170
Jacquard loom, 53, 73, 124
Jameson, Fredric, 120, 200, 252
Joler, Vladan, 236

Index   261

Jones, Matthew L., 11–12, 58–9, 223, 242

Kaldrack, Irina, 242
Kalindi, Vora, 8, 229
Kang, Minsoo, 52
Kant, Immanuel, 32, 34, 138–9
Katz, Yarden, 10, 12, 24, 229
Keller, Evelyn Fox, 13, 139
Killen, Andreas, 179
Kittler, Friedrich, 11, 155
Kline, Ronald R., 140, 154
Klüver, Heinrich, 163, 166
Knight, Bruce, 216
Knuth, Donald, 28–9, 155
Koffka, Kurt, 162, 170
Köhler, Wolfgang, 162–4, 166, 169–73, 178
Krämer, Sybille, 31, 43
Krizhevsky, Alex, 14
Kurenkov, Andrey, 215

labour, 2–9, 12–14, 18–19, 21, 25, 32, 35, 39, 47, 51–5, 59, 62–3, 65, 67, 74–7, 79–80, 82–3, 85–90, 93–4, 98–100, 102–10, 112–30, 135, 154, 158–9, 186, 191, 202–20, 237–40, 243–53
  abstract, 99, 119, 124–5, 203–4, 239
  automation of, 11, 16, 22, 64, 68, 135, 154, 230, 232–3, 236, 239, 245–8, 252
  calculation, principle of, 59, 62. *See also* Babbage principle
  cognitive, 1, 110
  concrete, 203
  division of, 4, 6–7, 17–18, 51, 54, 57, 59, 62–4, 67, 76, 78, 82–3, 88–90, 103–8, 116, 118–19, 122, 124–7, 190–1, 202, 204, 239, 243, 248, 251
  immaterial, 103
  knowledge, 103
  as logic, 3
  manual, 16–17, 58, 78, 80, 83–4, 88–90, 93, 103, 113, 201, 233
  mental, 16–17, 21, 25, 48, 52–3, 57, 59, 63–4, 66, 76, 83–4, 87–8, 93–4, 98–100, 108, 114, 116, 119, 130, 139, 197–8, 233–4, 246; demonisation of, 87
  of perception; of supervision; visual, 4, 17, 233
  social, 1, 6, 103, 110, 113, 129, 240
labour theory
  of automation, 17, 59, 63, 65, 104–5, 116, 120
  knowledge theory of labour, 83–8
  of knowledge, 88, 119
  of machine intelligence, 120
  of value, 6
Ladewig, Rebekka, 2
Lange, Oskar, 186, 199–201
Lardner, Dionysius, 74
Lashley, Karl, 163, 179, 188
Latour, Bruno, 117
Lazzarato, Maurizio, 101
Lebovic, Nitzan, 179
LeCun, Yann, 178, 216
Lee, Min Kyung, 251
Lefebvre, Henri, 241–2
Lefèvre, Wolfgang, 24, 39–40, 43, 91, 120, 255
Leibniz, Gottfried Wilhelm, 27, 42–3, 45–6, 58, 69, 166, 241
LeNet (convolutional neural network), 216
Leroi-Gourhan, André, 25, 242
Lettvin, Jerome Y., 173, 175
Lewin, Kurt, 241–2
Libbrecht, Elizabeth, 15
Light, Jennifer, 73
Logic Theorist, 31
Lovelace, Ada, 45, 54, 68–73
Lovelock, James, E., 138
Luchins, Abraham, 171
Luchins, Edith, 171
Luhmann, Niklas, 158
Lynch, Ben, 212

Mach, Ernst, 193
machine learning, 9, 12, 15, 19–20, 27–8, 47–8, 73, 94, 134–8, 148, 151, 155, 160, 163–4, 166, 176, 183, 191, 196, 212, 222–3, 226, 232, 234–6, 238, 243, 245, 247–9, 252
Machinery Question, 17, 63, 76–89, 93–4, 98, 100, 103, 114, 116, 122

Macho, Thomas, 30–1
MacKenzie, Donald, 101, 107
Maier, Charles S., 5
Malm, Andreas, 122–3
Marazzi, Christian, 100
Marcuse, Herbert, 102, 158
Marx, Karl, 1, 6, 18–19, 67, 75, 78,
    82–6, 88–9, 93–120, 125–6,
    190–1, 201–3, 233, 238–9
Mason, Paul, 101
master (factory's master) 1, 4–6
    eye of the, 4–6
Maturana, Humberto, 173
Mazières, Antoine, 14
McCarthy, John, 15, 134, 155, 163–4
McCulloch, Warren, 19–20, 134–7,
    148, 153–5, 161–75, 177, 181,
    183, 191, 193, 196, 212, 218,
    221, 246
McNaughton, Robert, 47
McQuillan, Dan, 252,
means of production, 82, 92, 107,
    113, 124, 127–8, 154, 239
Mechanical Turk, 81
    Amazon, 8, 9
Medina, Eden, 149
Menabrea, Luigi, 54, 69
Mezzadra, Sandro, 6, 119
Milbanke, Anne Isabella, 69
mindful hand, 2, 59, 89–90
Minsky, Marvin, 151, 216, 221, 230–2
Mirowski, Philip, 184, 187, 190
Mirzoeff, Nicholas, 4
Mises, Ludwig von, 186
Monakow, Constantin von, 179, 184,
    187–8
Moore, Doris L., 69
Morse, Samuel, 137
Mumford, Lewis, 32
Murray, Albert, viii, 208–9
Musk, Elon, 3
Musto, Marcello, 101

Negri, Antonio, 101–2, 120, 204
Neilson, Brett, 6, 119
Neisser, Ulric, 162, 194
neocognitron (convolutional neural
    network), 216
neural networks, 2–3, 14–15, 17,
    19–21, 27, 47, 53, 134–7, 141,
    143, 146, 148–52, 154, 159,
    162–6, 168, 170, 175–6, 178–9,
    183, 185–6, 188, 191–6, 198,
    204–7, 209, 211–12, 215–21,
    223, 226–7, 229–32, 234–5, 241,
    244–6, 249
    convolutional, 14, 15, 175–6, 231,
    235
neural telegraph, 133, 137
Newell, Allen, 146
Newton, Isaac, 27, 69, 92
Neyman, Jerzy, 229
Nicolaus, Martin, 101
Nik-Khah, Edward, 184, 187, 190
Nilsson, Nils John, 9, 163
Noble, David, 239
Noble, Safiya Umoja, 10

O'Connor, Sarah, 251
O'Neil, Cathy, 10
Olenina, Ana Hedberg, 5
Olivetti
    Elea 9003, 126
    factory, 126–7
Omodeo, Pietro Daniel, 13
OpenGraph protocol, 242
Otis, Laura, 133, 137
Owen, Robert, 97–8, 108
Owenism, 84–5, 110

Panzieri, Raniero, 101, 160
Papert, Seymour, 151, 216, 230–2
Parisi, Luciana, 248
Pascal, Blaise, 45, 58
Pask, Gordon, 147–8
Pasquale, Frank, 252
pattern, 8, 14, 24, 53, 73, 124, 144–5,
    147, 150, 163, 170, 174, 180,
    191, 193–6, 203, 210, 212,
    216–17, 223–5, 230, 241, 243,
    245, 247, 251
    recognition, 14–15, 17, 20, 47, 146,
    148, 150, 161–5, 168, 174, 180,
    182–3, 185, 193–6, 209, 213,
    216, 218, 222–4, 227, 233–5,
    240, 245
Pavoni, Andrea, 243
Pearson, Egon, 229
Penn, Jonathan, 224–5
Pentland, Alex, 243

Index 263

perceptron, 8, 15, 19–21, 143, 150, 162, 179, 183, 188, 192, 198, 222–7, 229–35, 244–7
   Mark I Perceptron, 150, 210–13, 216, 222, 225, 231
   phonoperceptron, 207
   photoperceptron, 207
   radioperceptrons, 209
Phillips, Leigh, 200, 252
Piaget, Jean, 35–8, 202
Pickering, Andrew, 92
Pinheiro dos Santos, Lúcio Alberto, 242
Pitts, Walter, 19–20, 134–7, 148, 153–5, 161–75, 177, 181, 183, 191, 193, 196, 212, 218, 221, 246
Plofker, Kim, 25
Polanyi, Michael, 189
Pollock, Frederick, 53
Prajapati, 24–5
Priestley, Mark, 67, 74, 165–6
Prigogine, Ilya, 138
Proudhon, Pierre-Joseph, 89, 105
Pryor, Nicholas, 207, 210, 216
punched card, 11, 53–4, 124, 193, 210
purusha, 25

Rabinbach, Anson, 79, 88
Ramasubramanian, K., 24
Rashevsky, Nicolas, 221
Reigeluth, Tyler, 235
Reisch, Gregor, 42
relations of production, 78, 82, 107, 154, 203, 239
relative surplus value, 104–5, 115–16, 118, 239
Renn, Jürgen, 13, 25, 48, 66, 202
ResNet-50 (convolutional neural network), 231
Ricardo, David, 80, 84
Riese, Walther, 187
Rishi, 25
Roberts, Lissa, 2, 89–90
Robinson, Cedric, 78
Röhle, Theo, 242
Rose, Hilary, 13
Rosenblatt, Frank, 15, 19–20, 141, 143, 148, 150–1, 164, 176, 179, 183, 188, 192, 197–8, 205–7, 209–11, 213–14, 216–27, 230–1, 234–5, 244, 247
Rosenblueth, Arturo, 139
Rossini, Manuela, 13
Rowe, William D., 163
Rozworski, Michal, 200, 251
Ryle, Gilbert, 189

Sacrobosco, Johannes de, 41
Sahlins, Marshall, 31
Salazar Sutil, Nicolás, 5
Samuel, Arthur, 20, 155
Savat, David, 242
Sawyer, Mellon, 10
Schaffer, Simon, 2, 9, 17, 52, 54, 59, 65–6, 73–4, 81, 89–90, 100, 104, 109, 120, 133, 153, 206, 227, 244
Schaupp, Simon, 241
Scheerer, Martin, 190
Schemmel, Matthias, 13
Schmidgen, Henning, 2, 13, 92, 99, 117, 122, 244, 246
Schrödinger, Erwin, 143
Schulz, Wilhelm, 98–9
science and technology history, 12, 27, 92, 106
self-driving vehicle, 1, 3, 235, 247
Selfridge, Oliver G., 151, 162, 194
Sellet, Frederic, 25, 242
Sennett, Richard, 2
Shanker, Stuart G., 158
Shannon, Claude, 46, 58, 136, 140, 155, 198, 241
Shapin, Steven, 82
Shaw, Cliff, 146
Shell, Marc, 201
Shulba Sutras, 24–5, 27
Silberman, Michael Six, 8
Simon, Herbert, 146, 153
Simondon, Gilbert, 121–3
Sloterdijk, Peter, 90
Smith, Adam, 4, 6, 55, 62, 86, 104, 111, 124, 239
Smith, Crosbie, 64, 75–6
Smith, David, 212
Smith, Hawkes, 82
Smith, Tony, 101
social constructivism, 12
social informatics, 12

sociotechnical history, 12
Sohn-Rethel, Alfred, 124, 201–2
Spearman, Charles, 21, 224, 228–9, 244
Sprenger, Florian, 3
Sriram, M. S., 26
Srnicek, Nick, 251
statistics, 6, 152, 220–3, 226, 228–9, 243–4
   automation of, 243, 248
   intuitive, 226, 229
   psychometrics, 21–2, 224–9, 235, 243–6
steam engine, 39, 52, 56, 75, 77, 91–2, 96, 107, 116, 122–4, 233, 239
   Watt's steam governor, 124
Staal, Frits, vii, 23, 25–6
Stark, Luke, 245
Stöcklein, Ansgar, 52
Suri, Siddharth, 8
Sutskever, Ilya, 14
Swade, Doron, 67, 71
Syenaciti, 24, 26

Taylor, Frederick Winslow, 5
Taylorism, 5, 104, 241
technopolitics, 253
Terman, Lewis, 21, 224
Terminator, The, 9
Terranova, Tiziana, 242
Teuber, Hans Lukas, 154
Teuscher, Christof, 134
Thales, 202
Thompson, Neil, 248
Thompson, William, 18, 75, 84–8, 93, 95, 98, 110–11, 114, 119
Thurstone, Louis Leon, 224–5, 228, 244
Tronti, Mario, 128–9, 159–60
Truitt, Elly Rachel, 9
Turing, Alan, 7–8, 19–20, 53, 58, 134, 139, 143–6, 152, 165–6, 170, 176, 198, 227, 241, 251

Turing machine, 19, 53, 143, 152, 165–6, 176, 198
Turkopticon, 8, 252
Turner, Fred, 156–7

Universal Constructor, 142–3
Ure, Andrew, 7–8
Uttley, Albert, 221

Vapnik, Vladimir, 222
Vercellone, Carlo, 101
Virilio, Paul, 162, 233
Virno, Paolo, 100–3
von Neumann, John, 19, 46, 58, 142–3, 153, 161, 163–6, 176–81, 188, 198, 211, 220–1
Vora, Kalindi, 8, 229
Voskhul, Adelheid, 9
Vygotsky, Lev, 235

Walter, Grey, 141
Wark, MacKenzie, 101
Watt, James, 123–4
Wendling, Amy E., 18
Wertheimer, Max, 32, 162, 169
Wiesel, Torsten, 175, 216
Wiggins, Chris, 11–12
Wing, Jeannette M., 23, 66
Winner, Langdon, 10, 13, 138
Winthrop-Young, Geoffrey, 11, 31, 155
Wise, Norton, 64, 75–6
Wolfe, Charles, 160
Wood, Alex, 251
Woodcock, Jamie, 14
Wynter, Sylvia, 252

Yovits, Marshall, 146–8, 151, 205–6
Yu Mingyi, 40

Zellini, Paolo, 25–7
Zilsel, Edgar, 90
Zuboff, Shoshana, 12
Zuse, Konrad, 143–4